T0269612

CAMBRIDGE LIBRARY COLLECTION

Books of enduring scholarly value

Earth Sciences

In the nineteenth century, geology emerged as a distinct academic discipline. It pointed the way towards the theory of evolution, as scientists including Gideon Mantell, Adam Sedgwick, Charles Lyell and Roderick Murchison began to use the evidence of minerals, rock formations and fossils to demonstrate that the earth was older by millions of years than the conventional, Bible-based wisdom had supposed. They argued convincingly that the climate, flora and fauna of the distant past could be deduced from geological evidence. Volcanic activity, the formation of mountains, and the action of glaciers and rivers, tides and ocean currents also became better understood. This series includes landmark publications by pioneers of the modern earth sciences, who advanced the scientific understanding of our planet and the processes by which it is constantly re-shaped.

A Treatise on the External, Chemical, and Physical Characters of Minerals

The renowned geologist Robert Jameson (1774–1854) held the chair of natural history at Edinburgh from 1804 until his death. A pupil of Gottlob Werner at Freiberg, he was in turn one of Charles Darwin's teachers. Originally a follower of Werner's influential theory of Neptunism to explain the formation of the earth's crust, he was later won over by the idea that the earth was formed by natural processes over geological time. He was a controversial writer, accused of bias towards those who shared his Wernerian sympathies, such as Cuvier, while attacking Playfair, Hutton and Lyell. This book, first published in 1805, of which the 1816 second edition is reissued here, gives physical descriptions of the minerals discussed in his three-volume System of Mineralogy (also reissued in this series). Dividing minerals into solid, friable and fluid types, he describes and gives the English, German, French and Latin names of each.

Cambridge University Press has long been a pioneer in the reissuing of out-of-print titles from its own backlist, producing digital reprints of books that are still sought after by scholars and students but could not be reprinted economically using traditional technology. The Cambridge Library Collection extends this activity to a wider range of books which are still of importance to researchers and professionals, either for the source material they contain, or as landmarks in the history of their academic discipline.

Drawing from the world-renowned collections in the Cambridge University Library and other partner libraries, and guided by the advice of experts in each subject area, Cambridge University Press is using state-of-the-art scanning machines in its own Printing House to capture the content of each book selected for inclusion. The files are processed to give a consistently clear, crisp image, and the books finished to the high quality standard for which the Press is recognised around the world. The latest print-on-demand technology ensures that the books will remain available indefinitely, and that orders for single or multiple copies can quickly be supplied.

The Cambridge Library Collection brings back to life books of enduring scholarly value (including out-of-copyright works originally issued by other publishers) across a wide range of disciplines in the humanities and social sciences and in science and technology.

A Treatise on the External, Chemical, and Physical Characters of Minerals

ROBERT JAMESON

CAMBRIDGE
UNIVERSITY PRESS

CAMBRIDGE
UNIVERSITY PRESS

University Printing House, Cambridge, CB2 8BS, United Kingdom

Cambridge University Press is part of the University of Cambridge.
It furthers the University's mission by disseminating knowledge in the pursuit of
education, learning and research at the highest international levels of excellence.

www.cambridge.org
Information on this title: www.cambridge.org/9781108084215

This edition first published 1816
This digitally printed version 2018

ISBN 978-1-108-08421-5 Paperback

A

TREATISE

ON THE

EXTERNAL,

CHEMICAL, AND PHYSICAL CHARACTERS

OF

MINERALS.

BY

ROBERT JAMESON,

REGIUS PROFESSOR OF NATURAL HISTORY,

AND

LECTURER ON MINERALOGY IN THE UNIVERSITY OF EDINBURGH,

&c. &c. &c.

SECOND EDITION.

EDINBURGH:

Printed by Neill & Company,

FOR ARCHIBALD CONSTABLE AND COMPANY, EDINBURGH; AND

LONGMAN, HURST, REES, ORME & BROWN, LONDON.

1816.

ADVERTISEMENT.

THIS Treatise contains a full account of the various External, Chemical, and Physical Characters employed in the descriptions of Minerals in my System of Mineralogy. It embraces not only the terminology of WER-NER, and other German naturalists, but also that of the celebrated HAUY; and to assist the Student, I have given, along with the English terms, also the German, French and Latin

The history of this branch of Mineralogy from the time of AGRICOLA, who published the first systematic arrangement of the Characters of Minerals, to that of the more perfect methods of WERNER and HAUY, would afford an opportunity of communicating much curious information ; but it is so extensive, that we must abandon it for the present, and rest satisfied with the following enumeration of the authors who have treated on this subject :

1. Agricola,

1. Agricola, in his work De Natura Fossilium. Basil, 1546, fol.
2. Chr. Aug. Hausen, Progr. ad solennia promotion. Magist. Leipsiæ, 1737. 4to.
3. Johan. Gottsch. Wallerii Mineral-riket. Holm, 1747. 8vo.
4. Frid. Aug. Cartheuseri Elementa Mineralogiæ. 1755. 8vo.
5. Gehler de Characteribus Fossilium Externis. 1757.
6. Valmont de Bomare, Mineralogie. A Paris, 1762. 8vo.
7. Caroli à Linné, Systema Naturæ. Holmiæ, 1768. 8vo. t. iii. p. 29. & 30. ; also C. à Linné, Amœnitates Academicæ, t. i. Dissert. de Crystallorum generatione ; Respond. Martino Kaehler. Holm, 1750. 8vo.
8. Joh. Thad. Peithnero, Erste grunde der Bergwerkswissenschaften, zweite Abhandlung, über Mineralogie. Prag. 1770. 8vo.
9. Hill, Fossils arranged according to their obvious Characters. London, 1771. 8vo.
10. Von den aüsserlichen Kennzichen der Fossilien abgefasst, von Abraham Gottlob Werner. Wien. 1774.
11. Walker's Delineatio Fossilium, in usus academicos Edinburgi. 1782.
12. Des Caracteres Exterieurs des Mineraux, ou reponse à cette question, Existe-t-il dans les substances du Regne Mineral des caracteres qu'on puisse regarder comme specifiques ; et au cas qu'il en existe, quel sont ces caracteres ? Par Romé de Lisle. A Paris, 1783.
13. Estner's Versuch einer Mineralogie. 1793.
14. Principes de Mineralogie, ou exposition succinte des Caractères Exterieures de Fossiles d'après les Leçons du Prof. Werner, augmentées d'additions manuscrites fournies par cet auteur. Par Vanberchem Berthoud et Struve. A Paris, 1794,—5. 8vo.
15. Tabulæ Synopticæ terminorum Systematis oryctognostici Werneriani, Latino, Danicæ, et Germanicæ, editæ a Gregorio Wad. Hafniæ, 1798, fol.

16. Weaver's

16. Weaver's Translation of Werner's Treatise on the External Characters of Minerals. 1800.

17. Hauy, Traité de Mineralogie, t. i. & ii. 1801.

18. Traité Elementaire de Mineralogie, suivant les Principes du Professeur Werner. Par Brochant, t. i. an. 9. (1802).

19. Handbok i Oryctognosien af G. M. Schwartz. Stockholm, 1803. 8vo.

20. Hausmann's Versuch eines entwurfs zu einer einleitung in die Oryctognosie. Braunschweig und Helmstadt, 1805. 8vo.

21. Leçons de Mineralogie, donnés au College de France. Par J. C. Delamethrie, t. i. 1811.

22. Hoffmann's Mineralogie, b. i. 1812.

CONTENTS

CONTENTS.

Regular

CHARACTERS of MINERALS.

THE characters of minerals are of different kinds, viz. *External, Chemical, Physical, Geognostical,* and *Geographical.*

1. *External Characters,*—are those which we discover by means of our senses, in the aggregation of minerals, and which have no reference to their relation to other bodies, or to chemical investigations.

2. *Chemical Characters,*—are those which are afforded by the complete analysis of the mineral, by trials with the various re-agents, the blowpipe, and the pyrometer.

3. *Physical Characters,*—are those physical phenomena which are exhibited by the mutual action of minerals and other bodies; such are the magnetic and electric properties exhibited by some minerals.

4. *Geognostical Characters,*—are those derived from various geognostic relations of minerals.

5. *Geographical Characters,*—are derived from the geographical distribution of minerals.

We shall first consider the External Characters, and then the others, in the order already mentioned.

A

External Characters of Minerals.

The External Characters of Minerals are either *generic* or *specific*. The Generic Characters are certain properties of minerals used as characters, without any reference to their differences, as colour, lustre, or weight. The differences among these properties form the Specific Characters, as adamantine lustre, and glassy or vitreous lustre. The generic characters are divided into *general* and *particular*: Under the first, are comprehended those that occur in all minerals, whether solid, friable, or fluid : under the second, those which occur only in particular classes of minerals. In the following tabular view, the External Characters are arranged nearly in a natural succession, and in the order in which they are employed in the descriptions of minerals.

A

A

TABULAR VIEW

OF THE

GENERIC EXTERNAL CHARACTERS

OF

MINERALS.

———◆———

GENERAL GENERIC EXTERNAL CHARACTERS.

1. COLOUR.

2. The Cohesion of the Particles, according to which minerals are distinguished into

Solid,	Friable,	Fluid

Particular generic characters of Solid minerals.	Particular generic characters of Friable minerals.	Particular generic characters of Fluid minerals.
External shape.	External shape.	
External surface.		
External lustre.		The lustre.
Lustre of the fracture.	The Lustre.	
The fracture.	Aspect of the particles.	
Shape of the fragments.		

(Left-hand bracketing: "Characters for the sight." — comprising "External aspect." { External shape, External surface, External lustre } and "Aspect of the fracture." { Lustre of the fracture, The fracture, Shape of the fragments })

Particular generic characters of Solid minerals.			Particular generic characters of Friable minerals.	Particular generic characters of Fluid minerals.
Characters for the sight.	Aspect of the distinct concretions.	Shape of the ditinct concretions. Surface of the distinct concretions. Lustre of the distinct concretions.		Transparency.
	General aspect	The transparency. The streak. The soiling.	The soiling.	Fluidity.
	Characters for touch.	The hardness. The tenacity. The frangibility. The flexibility. The adhesion to the tongue.	The friability.	
For the hearing.		The sound.		

Remaining general generic external characters.

For the touch. {
  3. The Unctuosity.
  4. The Coldness.
  5. The Weight.
For the smell.   6. The Smell.
For the taste.   7. The Taste.

TABULAR

5

TABULAR VIEW

OF THE

DIFFERENT GENERIC and SUBORDINATE SPECIFIC EXTER-
NAL CHARACTERS

OF

MINERALS.

GENERAL GENERIC EXTERNAL CHARACTERS.

I. COLOUR.

1. *The different Chief or Principal Colours and their varieties.*

A. WHITE. Weiss. Blanc. Albus.

a. *Snow-white.* Schneeweiss. Blanc de neige. Ni-
veo-albus.

b. *Reddish-white.* Röthlichweiss. Blanc rougeatre.
Rubescenti-albus.

c. *Yellowish-white.* Gelblichweiss. Blanc jaunatre,
Flavescenti-albus.

d. *Silver-white.* Silberweiss. Blanc d'argent. Ar-
genteo-albus.

e. *Greyish-white.* Graulich-weiss. Blanc grisatre.
Canescenti-albus.

f. *Greenish-white.* Grünlich-weiss. Blanc verdatre. Viridescenti-albus.

g. *Milk-white.* Milchweiss. Blanc de lait. Lac-
teo-albus.

h. *Tin-white.* Zinnweiss. Blanc d'étain. Stanneo-
albus.

B. GREY.

B. GREY. Grau. Gris. Griseus.

a. *Lead-grey.* Bleigrau. Gris de plomb. Plumbeo-griseus.

α. *Common-lead grey.* Gemeines bleigrau. Gris de plomb commun.

β. *Fresh lead-grey.* Frisches bleigrau. Gris de plomb fraiche.

γ. *Blackish lead-grey.* Schwärzlich bleigrau. Gris de plomb noiratre.

δ. *Whitish lead-grey.* Weisflich bleigrau. Gris de plomb albatre.

b. *Bluish-grey.* Bläulichgrau. Gris bleuatre. Cærulescenti-griseus.

c. *Pearl-grey.* Perlgrau. Gris de perle. Margaritino-griseus.

d. *Smoke-grey.* Rauchgrau. Gris de fumée. Fumoso-griseus.

e. *Greenish-grey.* Grünlichgrau. Gris verdatre. Viridescenti-griseus.

f. *Yellowish-grey.* Gelblichgrau. Gris jaunatre. Flavescenti-griseus.

g. *Ash-grey.* Aschgrau. Gris de cendre. Cinereogriseus.

h. *Steel-grey.* Stahlgrau. Cris d'acier. Chalybeogriseus.

C. BLACK. Schwarz. Noir. Niger.

a. *Greyish black.* Graulischwarz. Noir grisatre. Canescenti-niger.

b. *Iron black.* Eisenschwarz. Noir de fer. Ferreo-niger.

c. *Velvet black.* Sammet schwarz. Noir de velours. Atro-niger.

d. *Pitch black,* or *brownish black.* Bräunlichschwarz. Noir brunatre. Brunescenti-niger.

è. *Raven black,* or *greenish black.* Rabenschwarz oder grünlich schwarz. Noir verdatre. Viridescenti-niger.

f. Bluish

7

f. Bluish black. Bläulichschwarz. Noir bleuatre.
Cœrulescenti-niger.

D. BLUE. Blau. Bleu. Cœruleus.

  a. Blackish-blue. Schwarzlich blau. Bleu noiratre.
  b. Azure-blue. Lazurblau. Bleu d'azur. Azureo-cœruleus.
  c. Violet-blue. Veilchenblau. Bleu violet. Violaceo-cœruleus.
  d. Lavender-blue. Lavendelblau. Bleu de lavande, Lavendula-cœruleus.
  e. Plum-blue. Pflaumenblau. Bleu de prune. Pruneo-cœruleus.
  f. Berlin-blue, or *Prussian blue.* Berlinerblau. Bleu de Prusse. Berolino-cœruleus.
  g. Indigo-blue. Indigblau. Bleu d'indigo. Indico-cœruleus.
  h. Smalt-blue. Schmalteblau. Bleu de smalt. Smaltino-cœruleus.
  i. Duck-blue. Entenblau.
  k. Sky-blue. Himmelblau. Bleu de ciel. Cœlesti-cœruleus.

E. GREEN. Grün. Verd. Viridis.

  a. Verdigris-green. Spangrün. Verd de gris. Ærugineo-viridis.
  b. Celandine-green. Seladongrün. Verd celadon, ou de mer. Celadono-cœruleus.
  c. Mountain-green. Berggrün. Verd de montagne. Montano-viridis.
  d. Leek-green. Lauchgrün. Verd de poireau ou de prase. Prasino-viridis.
  e. Emerald-green. Schmaragdgrün. Verd emeraude, Smaragdino-viridis.
  f. Apple-green. Apfelgrün. Verd de pomme. Pomaceo-viridis.

g. Grass

8

g. Grass green. Grasgrün. Verd de pré. Grami-
neo-viridis.

h. Blackish green. Schärtzlichgrün. Verd noiratre.
Nigrescenti-viridis.

i. Pistachio-green. Pistaziengrün. Verd de pistache.
Pistacio-viridis.

k. Asparagus-green. Spargelgrün. Verd d'asperge.
Asparago viridis.

l. Olive-green. Olivengrün. Verd d'olive. Oliva-
ceo viridis.

m. Oil-green. Oelgrün. Verd d'huile. Oleario-
viridis.

n. Siskin-green. Zeisiggrün. Verd serin. Acan-
thino viridis.

F. YELLOW. Gelb. Jaune. Flavus.

a. Sulphur-yellow. Schwefelgelb. Jaune de soufre. Sulphureo-flavus.

b. Brass-yellow. Messinggelb. Jaune de laiton.
Orichalceo flavus.

c. Straw-yellow. Strohgelb. Jaune de paille. Stra-
mineo-flavus.

d. Bronze-yellow. Speissgelb. Jaune de bronze.
Æneo-flavus.

e. Wax-yellow. Wachsgelb. Jaune de cire. Ce-
reo-flavus.

f. Honey-yellow. Honiggelb. Jaune de miel. Mel-
leo-flavus.

g. Lemon-yellow. Citrongelb. Jaune de citron.
Citrino-flavus.

h. Gold-yellow. Goldgelb. Jaune d'or. Aureo-
flavus.

i. Ochre-yellow. Ockergelb. Jaune d'ochre. Ochra-
ceo-flavus.

k. Wine-yellow. Weingelb. Jaune de vin. Vineo-
flavus.

l. Pea-yellow. Erosengelb.

l. Cream,

9

l. Cream, or Isabella yellow. Isabelgelb. Jaune
isabelle. Isabelino-flavus.

m. Orange yellow. Oraniengelb, oder Pommeranz-
gelb. Jaune d'orange. Aurantio-flavus.

G. Red. Roth. Rouge. Ruber.

a. Aurora, or morning red. Morgenroth. Rouge
d'aurore Aurorco-ruber.

b. Hyacinth red. Hyazinthroth. Rouge d'hya-
cinthe ou ponceau. Hyacinthino-ruber.

c. Tile red. Ziegelroth. Rouge de brique. Lateri-
tio-ruber.

d. Scarlet red. Scharlachroth. Rouge ecarlate. Scar-
latino-ruber.

e. Blood red. Blutroth. Rouge de sang. Sangui-
neo-ruber.

f. Copper red. Kupferroth. Rouge de cuivre.
Cupreo-ruber.

g. Flesh red. Fleischroth. Rouge de chair. Car-
neo-ruber.

h. Carmine red. Karminroth. Rouge de carmin.
Carminco-ruber.

i. Cochineal red. Koschenillroth. Rouge de coche-
nille. Coccineo-ruber.

k. Crimson red. Kermesinroth. Rouge cramoisi.
Carmesino-ruber.

l. Rose red. Rosenroth. Rouge rose. Roseo-ru-
ber.

m. Peach-blossom red. Pfirsichbluthroth. Rouge de
fleurs de pecher. Persicino ruber.

n. Columbine red. Kolumbinroth. Rouge colum-
bin. Columbino-ruber.

o. Cherry red. Kirschroth. Rouge cerise. Cera-
sino-ruber.

p. Brownish red. Braunlichroth. Rouge brunatre.
Brunescenti-ruber.

B   H. Brown,

H. Brown. Braun. Brun. Brunus.

a. Reddish brown. Röthlichbraun. Brun rougeatre. Rubescenti-brunus.

b. Clove brown. Nelkenbraun. Brun de gerofles.
Caryophyllino-brunus.

c. Hair brown. Haarbraun. Brun de cheveux.
Capillari-brunus.

d. Broccoli brown. Kohlbraun. Brun de chou.
Brassicino-brunus.

e. Chesnut brown. Kastanienbraun. Brun de cha-
taigne. Castaneo-brunus.

f. Yellowish brown. Gelblichbraun. Brun jaunatre.
Flavescenti-brunus.

g. Pinchbeck brown. Tombackbraun. Brun de tom-
bac. Tombacino-brunus.

h. Wood brown. Holzbraun. Brun de bois. Lig-
neo-brunus.

i. Liver brown. Leberbraun. Brun de foie. Hepa-
tico-brunus.

k. Blackish brown. Schwärzlichbraun. Brun noiratre. Nigrescenti-brunus.

2. *The Intensity of the Colours.* Die höhe der farben. L'intensité des couleurs. Vis colorum.

A. Dark. Dunkel. Foncée. Obscurus.
B. Deep. Hoch. Relevée. Eminens.
C. Light. Lichte. Claire. Clarus.
D. Pale. Blass. Pâle. Pallidus.

3. *The delineations, or patterns, formed by the colours.* Die
Farbenzeichnung. Desseins de couleurs. Pictura colorum.

A. Dotted. Punctirt. Pointillé. Punctati.
B. Spotted. Gefleckt. Tacheté. Maculati.
C. Clouded. Gewolkt. Nuagé. Nubiformis.

D. Flamed.

D. FLAMED. Geflammt. Flambé. Flammei.
E. STRIPED. Gestreift. Rubanné. Fasciati.
  a. Straight. Gerade. Zonis rectis.
  b. Ring-shaped. Ringförmig. Annulaire. Zonis concentricis notati.
F. VEINED. Geadert. Veiné. Venati.
G. DENDRITIC. Baumförmig. Dendritique. Dendritici.
H. RUINIFORM. Ruinenförmig. Ruiniforme. Ruinæformes.

4. *The tarnished colours.* Angelaufenen farben. Couleurs superficielles. Colores superficiales,—*are distinguished,*

A. *According to their origin.* Enstehung. Origine. Origo.
  a. In the bosom of the earth. Sogleich auf der lagerstätte.
  b. On the exposition of the recent fracture to the action of the air. Bei oder auf jedesmaligen frsichen bruche.

B. *According to the kind.* Nach der art der farben. D'apres leur variations. Quoad aspectum.
  a. Merely simple. Einfache. Simples. Simplices.
    α. Grey. Grau.
    β. Black. Schwarz
    γ. Brown. Braun.
    δ. Reddish. Röthlich.
  b. Many party coloured together, (variegated). Mehrere zugleich. Bigarées. Variegati.
    α. Pavonine, or Peacock-tail. Pfauenschweifig. Queue de pavon. Pavonaceus.
    β. Iridescent, or Rainbow. Regenbogenfarbig. Iris. Iridei.
    γ. Columbine, or Pigeon-neck. Taubenhälsig. Gorge de pigeon. Columbinii.

                        δ. Tempered

3. Tempered steel coloured.   Gehärtete stahlfarben. Acier trempé.   Chalibei.

5. *The play of the colours.*   Farbenspiel.   Jeu de couleurs. Lusus colorum.

6. *The changeability of the colours.*   Die Farbenwandlung. La mutabilité des couleurs.   Variatio colorum.

   A. *On the surface, (observed by looking in different directions on the mineral.)*   Auf der oberfläche beim daraufschen.   A la surface.   In superficie.

   B. *Internally, (by looking through it).*   Inwending beim durchschen.   A l'interieur.   Intus.

7. *The Iridescence.*   Das irisiren.   *Observed by*

   A. *Looking on the mineral.*

   B. *Looking through it.*

8. *The permanent alteration of the colours.*   Die Farbenveränderung.   Alteration des couleurs.   Mutatio colorum.

II. THE COHESION OF THE PARTICLES.   Die Zusammenhang der Theile.   Cohesion.   Cohærentia partium.

1. *Solid, in general.*   Feste im allgemeinen.

   A. *Solid, in a stricter sense.*

   B. *Friable.*   Zerreibliche.

2. *Fluid.*

The remaining Generic Characters are placed at the conclusion of this Tabular view, that is, immediately after the particular Generic Characters.

2. Particular *external shape.* Besondere aussere gestalten. Formes exterieures imitatives. Figuræ externæ singulares.

A. *Longish.* Längliche. Alongées. Longiusculæ.

a. *Dentiform.* Zähnig. Dentiforme. Dentiformis.

b. *Filiform.* Drathförmig. Filiforme. Filiformis.

c. *Capillary.* Haarförmig. Capillaire. Capillaris.

d. *Reticulated.* Gestrickt. Tricoté ou en reseau. Retiformis.

e. *Dendritic.* Baumförmig. Dendritiforme. Dendritica.

f. *Coralliform or coralloidal* Zackig. Coralliforme. Coralliformis.

g. *Stalactitic.* Tropfsteinartig. Stalactiforme. Stalactica.

h. *Cylindrical.* Röhrenförmig. Cylindrique. Tubulosa.

i. *Tubiform.* Pfeifenröhrig. Tubiforme. Fistulosa.

k. *Claviform.* Kolbenförmig. Claviforme. Claviformis.

l. *Fruticose.* Staudenförmig. En buissons. Fruticosa.

B. *Roundish.* Runde, Rondes. Rotundæ.

a. *Globular.* Kuglich. Globuleuse. Globulosa.

α. *Perfect globular or spherical.* Sphœrish. Spherique. Sphærica.

β. *Imperfect globular.* Unvolkommen kuglich. Spherique imperfait.

γ. *Ovoidal or eliptical.* Elliptisch. Ovoide ou elliptique. Elliptica.

δ. *Spheroidal.* Spheroidisch. Spherique applati ou spheroidal. Sphæroidea.

ε. *Amygdaloid.* Mandelförmig. Amygdaliforme. Amygdaloidea.

b. *Botryoidal.*

b. Botryoidal. Traubich. Uviformes. Uvæformis.

c. Reniform. Nierförmig. Reniforme. Reniformis.

d. Tuberose. Knollig. Bulbeux ou tuberculeux. Tuberosa.

e. Fused-like. Geflossen. Coulée. Liquata vel fusa.

C. *Flat.* Platte. Plattes. Planæ.

    a. Specular. Spieglich. Speculaire ou miroiréc. Specularis.

    b. In leaves. In blättchen. En feuilles *ou* bractées. Bracteata.

D. *Cavernous.* Vertiefte. Creuses. Excavata.

    a. Cellular. Zellich. Cellulaire. Cellulosa.

       α. Straight or angulo-cellular. Geradzellig.

         1. *Hexagonal.* Sechsseitig.

         2. *Polygonal.* Viellseitig.

       β. Circulo-cellular. Rundzellich.

         1. *Parallel.* Gleichlaufend.

         2. *Spongiform.* Schwamförmig.

         3. *Indeterminate.* Unbestimmt.

         4. *Double.* Doppelt.

    b. Impressed. Mit eindrücken. Avec des impreintes. Impressa.

    a. *With impressions of crystals.*

       α. Cubical. Würflichen. Cubiques. Vestigiis cubicis.

       β. Pyramidal. Piramidalen. Pyramidales. Pyramidalibus.

       γ. Tabular. Tafelartagen. Tabuliformis. Tabulæformibus.

    b. *With impressions of particular external shapes.*

       α. Conical. Kegelförmigen. Coniques. Conicis.

       β. Globular. Kuglichen. Spheriques. Globosis.

       γ. Reniform. Nierformig. Reniform.

    c. Perforated.

c. *Perforated.* Durchlöchert. Criblé. Perforata.

d. *Corroded.* Zerfressen. Carié. Corrosa.

e. *Amorphous.* Ungestaltet. Informe. Monstruosa.

f. *Vesicular.* Blasig. Bulleuse. Bellulosa.

E. *Entangled.* Verworren. Emmelées. Implicata.

a. *Ramose.* Astig. Rameuse. Ramosa.

3. REGULAR *External Shape, or Crystallization.* Regelmässige aussere Gestalten. Formes exterieures regulières, ou crystallisations. Figuræ externæ regulares seu crystallisationes.

A. *The genuineness.* Die wesentlichkeit. Essentialité. Essentialitas : *according to which, crystals are either,*

a. *True.* Wesentliche. Vrais crystaux. Vera crystallisatio,—*or*

b. *Supposititious.* Aftercrystalle. Pseudo crystaux. Pseudo-crystalli.

B. *The shape.* Die Gestalt. Forme des cristaux. Figura crystallorum.

a. *Which is made up of*

α. *Planes.* Flächen. Faces. Plana.

β. *Edges.* Kanten. Bords. Margines.

γ. *Angles.* Ecken. Angles. Apices, *and*

b. *In which is to be observed,*

α. *The fundamental figure.* Die grundgestalt. Forme principale ou dominante. Figura fundamentalis.

(i.) *The parts of which are,*

1. *Planes, either*

A. *Lateral.* Seitenflächen. Faces laterales. Plana lateralia, *or*

B. *Terminal.* Endflächen. Faces terminales. Plana terminalia.

2. *Edges.* Kanten, *either*

C

A. *Lateral.*

18

A. *Lateral.* Seitenkanten. Bords lateraux. Margines laterales, *or*

B. *Terminal.* Endkanten. Bords terminaux. Margines terminales, *and*

3. *Angles.* Ecken.

(ii.) *The kinds of fundamental figure, which are*

1. *The icosahedron.* Icosaeder. Icosaedre. Icosaedrum.
2. *The dodecahedron.* Dodecaeder. Dodecaedre. Dodecaedrum.
3. *The hexahedron.* Hexaeder. Exaedre. Hexaedrum.
4. *The prism.* Säule. Prisme. Prisma.
5. *The pyramid.* Pyramide. Pyramide. Pyramis.
6. *The table.* Tafel. Table. Tabula.
7. *The lens.* Linse. Lentille. Lens.

(iii.) *The varieties of each kind of fundamental figure in particular, according to*

1. *Simplicity.* Einfacheit. Simplicité. Simplicitas : *which distinction is, however, eonfined to the pyramid, as occurring either*
   A. *Single.* Einfach. Simple. Simplex, *which is either*
      a. *Erect.* Rechts. Droite. Erecta.
      b. *Inverted.* Verkehrt. Renversée. Inversa, *or*
   B. *Double.* Doppelt. Double. Duplex,—*and then*
      a. *The lateral planes of the one pyramid set on the lateral planes of the other, either*
         α. *Streight.* Gerade. Droite, *or*
         β. *Oblique.* Schief. Biais, *or*
      b. *On the lateral edges of the other.*

2. *Numbers of the planes ; here we have to observe*
   A. *The species of the planes.* Art der fläehen. Espece des faces, as*

α. *In*

α. *In the prism and pyramid the lateral planes are different, and*

b. *In the tables the terminal planes.*

B. *The number of them, according to which they may be, either*

  a. Trihedral, or three-sided. Dreisetig. Trilatere.

  b. Tetrahedral, or four-sided. Vierseitig. Quadrilatere.

  c. Hexahedral, or six-sided. Sechseitig. Sextilatere.

  d. Octahedral, or eight-sided. Achtseitig. Octolatere.

3. *Proportional size of the planes to one another.* Verhältniss der flächen in ansehung der grösse zu einander. Grandeur des faces relativement les unes aux autres. Proportio planorum respectu magnitudinis.

A. *Equilateral.* Gleichseitig. Faces egales. Plana æqualis.

B. *With unequal planes.* Ungleichseitig. Faces inegales. Plana inæqualia: *either*

  a. Indeterminately unequal. Unbestimmt. Irregulierement inegales, *or*

  b. Determinately. Bestimmt. Regulierement inegales,—*which are*

    α. Alternately broad and narrow. Abwechselend breitere und schmälere. Alternativement larges et etroites.

    β. Two opposite planes broader. Zwei gegenüberstehende breitere seitenflächen. Deux faces larges opposées.

    γ. Two opposite planes narrower. Zwei gegenüberstehende schmälere seitenflächen. Deux faces etroites opposées.

4. *The* Direction *of the faces or the planes.* Richtung der
flächen. Forme des faces. Directio planorum, *which is*

A. *Rectilinear or straight.* Geradflächig. Plane. Rectiplana, *or*

B. *Curvilinear.* Krummflächig. Courbé. Curviplana.—*These
differ partly by*

   a. The position of the curvature. Nach der lage der krum-
mung. Position de la courbure. Situs,—*being*

   α. *Concave.* Einwärts gekrümmt. Concave. Concava.
   β. *Convex.* Auswärts gekrümmt. Convexe. Convexa.
   γ. *Concavo-convex.* Ein und auswärts gekrümmt,—
   and partly by

   b. The shape. Nach der gestalt. Espece de courbure.
Figura,—*which is either*

   α. *Spherical.* Sphärisch.
   β. *Cylindrical.* Cylindrisch.

   (1.) *The convexity parallel with the length or breadth
of the sides.* Die convexität mit den seitenflächen gleichlaufend,—*or*

   (2.) *The convexity parallel with the diagonal.* Die
convexität mit den diagonale gleichlaufend.

   γ. *Conical* Conisch. Conique. Conica.

5. *The* Angles *under which the planes meet.* Winkeln, unter
welchen die flächen zusammenstossen. Anglesdes faces
entre elles. Quantitas angulorum:—*these are either*

   A. *The lateral edges.* Seitenkantenwinkel. Bords lateraux. Anguli marginales laterales,—*which are*

   a. Equiangular. Gleichwinklich. Equiangles. Æ-
quales.
   b. Unequiangular. Verschiedenwinklich. Inegaux.
Diversi: *or*

            B. *The*

B. *The terminal edges.* Endkantenwinkel. Bords terminaux. Anguli marginales terminales,—*which are*

a. *Rectangular.* Recht. Rectangules. Recti, *or*
b. *Obliquangular.* Schief, *and this*
α. *Parallel oblique.* Gleichlaufend schief.
β. *Alternate oblique.* Abwechselnd schief.

c. *The summit angle.* Endspitzenwinkel. Angle solide du sommet ou pointe. Anguli apicis,—*which may be*

a. *Uncommonly obtuse.* Ausserst stumpfe oder flache. Extremement obtus.
b. *Very obtuse.* Sehr flach oder stumpfe. Tres obtus.
c. *Obtuse.* Flach oder stumpfe. Obtus.
d. *Rather obtuse.* Ein wenig flach oder stumpf. Un peu obtus.
e. *Rectangular.* Rechtwinklich. Rectangulaire.
f. *Rather acute.* Ein wenig spitzig oder scharf. Un peu pointu.
g. *Acute.* Spitzig oder scharf. Pointu.
h. *Very acute.* Sehr spitzig oder scharf. Tres pointu.
i. *Uncommonly acute.* Aussert scharfe oder spitzig. Extremement aigus.

6. *The* Magnitude *of the angles.* Grosse der winkel. Valeur des angles.

7. Plenitude *of the crystals.* Völle des crystalls. Plenitude des cristaux. Plenitudo crystallorum,—*either*

A. *Full.* Voll. Plein. Plenæ.
B. *Excavated at the extremities.* Ausgehölt an den enden. Creusc à l'extremité. Terminis excavatæ.
C. *Hollow.* Hohl. Vuide. Cavæ.

(iv.) *The alterations of the fundamental figure take place by*

I. *The Truncation.* Abstumpfung. Troncature. Truncatura.—*Here we have to consider*

1. *The parts of the truncation.* Die theile der abstumpfung. Parties de la troncature. Partes truncaturæ.—*These are*

A. *The planes of the truncation.* Abstumpfungsflächen. Faces de la troncature. Plana truncaturæ.

B. *The edges of the truncation.* Abstumpfungskanten. Bords de la troncature. Margines truncati.

C. *The angles of the truncation.* Abstumpfungsecken. Angles de la tronc. Apices truncaturæ.

2. *The determination of the truncation.* Die bestimmung der abstumpfung. Determination de la troncature. Determinatio truncaturæ,—*which relates to*

A. *The placing of the truncation, or its situation.* Ort. Place de la troncature. Locus.

a. *On the edges.* An den kanten. Aux bords. Marginibus.

b. *On the angles.* An den ecken. Aux coins. Apicibus.

B. *Magnitude of the truncation.* Stärke oder grösse. Grandeur de la troncation. Magnitudo.

a. *Deep.* Stark. Forte. Multum truncatum.

b. *Slight.* Schwach. **Legere.** Pàrùm truncatum.

C. *The setting on or application of the truncation.* Aufsetzung. Position relative de la troncature. Applicatio planorum.

a. *Straight.*

  a. Straight.   Gerade.   Droite.   Recta applicata.

  b. Oblique.   Schief.   De biais.   Obliquè appli-
    cata.

D. *The direction of the truncating planes.*   Die richtung
  der abstumpfungs flache.   Forme de la troncature.
  Directio planorum,—*which are*

  a. Rectilinear.   Geradflächig.   Plane.   Rectiplana.

  b. Curvilinear, or roundeed of.   Krumflächig.   Courbe.
    Curviplana.

II. *The Bevelment or Cuneature.*   Zuscharfung.   Bisellement.
  Acumen.   *Here we have to consider :*

  1. *The parts of the bevelment.*   Theile der zuschärfung.
   Parties du bisellement.   Partes acuminis.   *These are,*

   A. *The planes of the bevelment.*   Die zuschärfungflä-
    chen.   Faces.   Plana acuminis.

   B. *The edges of the bevelment.*   Die kanten der zu-
    schärfung.   Bords.   Margines.

    a. The proper edge.   Die eigentliche zuschärfungs-
    kante.   Bord formé par les deux faces du biselle-
    ment.   Proprii acuminis ; *and*

    b. The edges formed by the bevelling and lateral planes.
    Die kanten zwischen den zuschärfungs-und-seiten
    flächen.   Bord formé par les faces du bisellement
    et les autres.   Margines inter planem acuminis
    et lateralia.

   C. *The angles of the bevelment.*   Die zuschärfungsecken.
    Coins.   Apicis acuminis.

  2. *The determination of the bevelment.*   Bestimmung.   Determination du bisellement.   Determinatio acuminis.
  Here we have to observe,

   A. *The situation.*   Ort.   Place du bisellement.   Locus.

   a. On the terminal planes.   An den endflächen.   Aux
    faces terminales.   Planis terminalibus.

                  b. On

b. On the edges. An den kanten. Aux bords. Marginibus : *and*

c. On the angles. An den ecken. Aux coins. Apicibus.

B. *The magnitude.* Die stärke. Grandeur du bisellement. Magnitudo. *According to which, it is either*

a. Deep. Stark. Fort. Multum, *or*

b. Slight. Schwach. Legere. Parvum.

C. *The angle.* Der winkel. Bord propre ou angle simple formé par les faces du bisellement. Angulus acuminis.

a. Obtuse. Flach. Obtus. Obtusus.

b. Rectangular. Rechtwinklich. Rectangulaire. Rectangulus.

c. Acute. Scharf. Aigu. Acutus.

D. *The uniformity.* Die fortdauer.

a. Uniform. Ungebrochen.

b. Broken. Gebrochen. Fractus.

α. Once broken. Einmal gebrochen.

β. Twice broken. Zweimal gebrochen.

E. *The application.* Die aufsetzung. Position relative du bisellement. Applicatio.

a. Of the bevelment itself. Die zuschärfung selbst. Position du bisellement. Acuminis ipsius. *Which is either*

α. Straight. Gerade. Droit. Recta, *or*

β. Oblique. Schief. De biais. Obliqua.

b. Of the planes. Flächen. Celle des biseaux. Planorum.

α. On the lateral planes. Auf die seitenflächen. Sur les faces laterales. Ad plana lateralia.

β. On the lateral edges. Sur les bords lateraux. Ad margines laterales.

III. *The*

III. *The Acumination.* Zuspitzung.

Here we have to consider,

1. *The parts of the acumination.* Die theile der zuspitzung. Parties du pointement. Partes mucronis : *which are*

A. *Acuminating planes.* Zuspitzunflächen. Faces. Plana.

B. *Edges of the acumination.* Zuspitzungskanten. Bords. Margines, *which are either*

a. *Acuminating edges.* Die eigentliche zuspitzungskanten. Bords du pointement même.

b. *Terminal edges of the acumination.* Die endkanten der zuspitzung. Bord terminal du pointement.

c. *Edges formed by the acuminating and lateral edges.* Die kanten, welche die zuspitzungsflächen mit den seitenflächen machen. Bords que les faces du pointement font avec les autres.

C. *The acuminating angles* Die zuspitzungsecken. Coins du pointement. Apices.

a. *The angles between the acuminating planes, and the lateral planes of the fundamental figure.*

b. *The terminal or summit angle.*

2. *The determining the acumination depends on observing,*

A. *The situation of it.* Ort. Place du pointment. Locus, *either*

a. *On the angles.* An den ecken. Aux coins. Apicibus, *or*

b. *On the extremities.* An den enden. Aux faces terminales. Terminis.

B. *The acuminating planes.* Die zuspitzunflachen. Faces du pointement. Plana.

a. *Their number.* Deren anzahl. Leur nombre. Numerus.

D   *b. Their*

b. Their proportional magnitude between themselves. Deren verhaltnissmässige grösse gegen einander. Grandeur relative entre elles. Magnitudo mutua.

c. Their shape. Deren gestalt. Leur contour. Figura, *either*

α. *Determinate.* Bestimmt. Regulier. Determinata, *or*
β. *Indeterminate.* Unbestimmt. Irreguliere. Indeterminata.

d. Their setting on. Die aufsetzung. Position. Applicatio, *either*

α. *On the lateral planes.* Auf den seitenflächen. Sur
les faces de la forme simple. Ad plana lateralia, *or*
β. *On the lateral edges.* Auf die seitenkanten. Sur
les bords de la forme simple. Ad margines laterales.

C. *The summit angle.* Der winkel der zuspitzung. Bord
du pointement. Angulus : *which is*

a. Obtuse. Flach. Obtus. Obtusus.
b. Rectangular. Rechtwinklich. Rectangulaire. Rec-
tus.
c. Acute. Scharfwinklich. Aigu. Acutus.

D. *The magnitude.* Die stärke. Grandeur du pointement.
Magnitudo : *according to which, crystals are*

a. Deeply acuminated. Stark. Fort. Multum mucro-
natum : *or*
b. Slightly acuminated. Schwach. Faible. Parum mu-
cronatum.

E. *The termination.* Die endigung. Terminaison du pointement. Terminatio : *as the acumination may terminate*

a. In a point In einen punct. Un point. In punctum :
or
b. In a line. En un line. Une ligne. In lineam.

IV. The

IV. The *Division* of the Planes.

1. *The number, as into two, three, four, or six compartments.*
2. *The direction of the dividing edges.*
    a. *In the direction of the diagonal.*
    b. *From the middle part of the plane towards the angles and edges.*

V. *Multiplied alterations.* Mehrfachen veränderung der grundgestalt : *which occur in certain crystals, and which are either*

1. *Co-ordinate.* Nebeneinandergesetzt, *or*
2. *Superimposed.* Ubereinandergesetzt.

For the more exactly determining a crystallization, may be adjoined the general determination of its planes ; and then

α. *The number of the planes in general, and of each species in particular ; and*
β. *The shape of each species of plane, must be given.*

Besides these, in describing a crystallization, the following may be observed and adjoined :

a. *The choice of different modes of describing one and the same crystallization.*

The principal or most essential form of a crystallization will be, however, determined

α. *By the larger planes,*
β. *By the greater regularity,*
γ. *By its most frequent occurrence,*
δ. *By its affinity with the other fuudamental forms of the same fossil.*
ε. *By the suitability and adaptation to the alterations which occur in the crystal suite or crystallization ; and*
ζ. *By the greater simplicity.*

b. *The*

b. The transitions which arise from thence,

    *α. That the new or alterating planes become gradually
larger, at the expence of certain previous planes,
which are at length wholly obliterated,*

    *β. By alterations taking place in the proportion of the
planes between themselves,*

    γ. By alteration of the angles.

    δ. By convexity, and

    ε. By aggregation.

*c. Obstacles which prevent, or at least render the exact de-
termination of certain crystals, difficult, are occasioned
by*

    α. Their obliquity. Verschobenseyn. L'allonge-
ment. Obliquitas planorum et angulorum.

    β. Their incorporation. Verwachsenseyn. L'Incor-
poration dans un fossil. Coalescentia.

    γ. Their being broken. Verbrochenseyn. Breches.
Ruptura, *and*

    δ. Their too great minuteness. Die zu grosse klein-
heit. La trop grand petitesse. Nimia parvitas.

C. *The Magnitude of the Crystals.* Die Grösse der Krystallen.

    *a. With regard to their absolute magnitude, crystals are
divided into,*

    α. Uncommonly large. Ungewöhnlich gross. Ex-
tremement grand. Eximiè grandes.

    β. Very large. Sehr gross. Tres grand. Pergrandes.

    γ. Large. Gross. Grand. Grandes.

    δ. Middle sized. Von mittlerer grösse. Moyenne
grandeur. Mediocriter grandes.

    ε. Small. Klein. Petit. Parvæ.

    ζ. Very small. Sehr klein. Tres petit. Minutæ.

    η. Microscopic. Ganz klein. Tout petit. Minu-
tissimæ.

                                             b. In

b. In describing the relative magnitude of crystals. Die relative grösse. La grandeur relative,—*the following terms are used:*

(*α.*) *In describing the prism.*

 a. *In regard to length,*

  aa. *Short or low.* Kurz oder niedrig. Court.

  bb. *Long or high.* Lang oder hoch. Long.

 b. *In regard to breadth and thickness.*

  aa. *Broad.* Breit. Large.

  bb. *Acicular.* Nadelförmig. Aciculaire.

  cc. *Capillary.* Haarförmig. Capillaire.

(*β.*) *In describing the pyramid.*

 a. *In regard to length.*

  aa. *Short or low.* Kurz oder niedrig. Court.

  bb. *Long or high.* Lang oder hoch. Long.

 b. *In regard to breadth and thickness.*

  aa. *Broad.* Breit. Large.

  bb. *Subulate.* Spiessig. Subulé.

(*γ.*) *In describing the table.*

 a. *In regard to length and breadth.*

  aa. *Longish.* Länglich. Long.

 b. *In regard to thickness.*

  aa. *Thick.* Dick. Epais.

  bb. *Thin.* Dünn. Mince.

(*δ.*) *Crystals, in which all the dimensions are alike, are named Tessular.*

D. *The*

D. *The Attachment of the Crystals.* Der zusammenhang der krystallen. Le grouppement ou l'adherence des crystaux entre eux. Aggregatio crystallorum. *According to which they may be either*

a. Solitary. Einzeln. Separés. Solitariæ; *and this again*

α. *Loose.* Lose. Isolé ou solitaire. Solutæ.

β. *Imbedded.* Eingewachen. Implanté. Innatæ, *or*

γ. *Superimposed.* Aufgewachsen. Superposé. Adnatæ.

b. Aggregated. Zusammengehäuft. Groupes aggregés. Connata, *either*

(α.) *A determinate number growing together in a determinate manner;*

1. *With respect to Number,*

   i. *Pair wise, (twin crystals.)* Zwillingscrystalle. Jumeaux. Gemellæ.

   ii. *Three together, (triple crystals.)* Drillingscrystalle. Jumeaux triples. Tergeminæ.

   iii. *Four together, (quadruple crystals.)* Vierlings crystalle. Crystaux quadruples.

2. *With regard to the manner of, their connection.* Zulammenfügung.

   i. *Intersecting one another.* Durcheinandergewachsen.

   ii. *Penetrating one another.* Ineinandergewachsen.

   iii. *Adhering to one another.* Aneinandergewachsen.

(β.) *Many together, but merely simply aggregated.* Einfach zusammengehäuft; *either*

   i. *On one another.* Aufeinander. Les uns sur les autres. Superimpositæ.

   ii. *Side*

31

ii. *Side by side.* Aneinander. Les uns à cotés
des autres. Adpositæ, *or*

iii. *Promiscuously.* Durcheinander gewachsen. Sans
ordre. Decussatæ.

(γ.) *Many together, doubly aggregated.* Mehrere doppelt zusamengehäuft. Plusieurs crystaux doublement aggregés. Plures dupliciter connatæ. *The
most remarkable are,*

*In longish Crys-
tals.*
{
  i. *Fascicular or scopiform.* Buchelförmig.
    En faisceau. Fasciculatim.
  ii. *Manipular or sheaf-like.* Garbenförmig.
  iii. *Columnar.* Stangeförmig. En barres.
  iv. *Pyramidal.* Pyramidal. En pyramides.
    Pyramidaliter.
}

v. *Bud-like.* Knospenförmig. En boutons.
Gemmæformiter.

*In tabular Crys-
tals.*
{
  vi. *Rose-like.* Rosenförmig. En rose. Rosæ-
    formiter.
  vii. *Amygdaloidal.* Mandelförmig. En aman-
    des. Amygdalorum instar.
}

*In roundish or
tessular Crys-
tals.*
{
  viii. *Globular.* Kuglich ou kugelförmig. En
    boule. Globosè.
  ix. *In rows.* Reihenförmig. En rayes. Or-
    dinatim.
}

x. *Scalarwise aggregated.* Treppenförmig.
en Escaliei.

4. EXTRANEOUS

4. EXTRANEOUS *external shape.* (PETRIFACTIONS.) Fremdartge aüssere gestalten; Versteinerungen.

A. *From the Animal Kingdom.*
  a. Of quadrupeds. Saugethieren.
  b. Of birds. Vögeln.
  c. Of amphibious animals. Amphibien.
  d. Of fishes. Fischen.
  e. Of insects. Insecten.
  f. Of shells. Schaalthieren, *as*
    * *Univalves.*
    i. *Belemnites.*
    ii. *Ammonites.*
    iii. *Turbinites.*
    iv. *Strombites, &c.*
    ** *Bivalves.*
    i. *Chamites.*
    ii. *Terebratulites.*
    iii. *Mytulites.*
    iv. *Gryphites.*
    v. *Ostracites, &c.*
    *** *Multivalves.*
    i. *Balanites, &c.*

  g. Of crustaceous animals, as echinites, asterites, &c.
  h. Of corals, as madreporites, reteporites, encrinites, entrochites, &c.

B. *From the Vegetable Kingdom,*
  a. Impressions of plants
  b. Petrified wood.

II. THE

II. *The External Surface.* Die aussere oberfläche. La surface exterieure. Superficies externa.

1. *Uneven.* Uneben. Inegale. Inæqualis.
2. *Granulated.* Gekörnt. Granulée. Granata.
3. *Rough.* Rauh. Apre. Aspera.
4. *Smooth.* Glatt. Lisse. Lævis.
5. *Streaked.* Gestreift. Striée. Striata.

A. *Simply streaked.* Einfach gestreift. Simplement striée, Simpliciter striata.

    a. Longitudinally. In die queere gestreift. En travers. Latitudinaliter.

    b. Transversely. In die lange gestreift.

    c. Diagonally. Diagonaliter. Diagonalement. Diagonaliter.

    d. Alternately. Abwechselnd gestreift. Rayée. Alternè.

B. *Doubly streaked.* Doppelt gestreift. Doublement striée. Dupliciter striata.

    a. Plumiformly. Federartig. En barbes de plumes. Pennatim.

    b. Reticularly. Gestrickt gestreift. En tricot. Reticulatim.

6. *Drusy.* Drusig. Drusique. Drusica.

III. *The External Lustre* Der aussere glanz. L'eclat exterieur. Nitor externus.

1. *The intensity of the lustre* Stärke des glanzes. Intensité ou degrés de l'eclat. Gradus nitoris.

Here we have to determine the following degrees:

A. *Splendent.* Starkglänzend. Tres eclatant Multum nitens.

E

B. *Shining*

B. *Shining.* Glänzend. Eclatant. Nitens.

C. *Glistening.* Wennigglänzend. Peu eclatant. Parum nitens.

D. *Glimmering.* Schimmernd. Brillant ou tremblant. Micans.

E. *Dull.* Mat. Mat. Nitoris expers.

2. *The sort of lustre.* Art des glanzes. Espece d'eclat. Species nitoris.

  A. *Metallic lustre.* Metallischer glanz. Eclat metallique. Nitor metallicus.

  B. *Common lustre.* Gemeiner glanz: *which is distinguished into*

    a. Semimetallic. Halbmetallischer glanz. Demimetallique. Semimetallicus.

    b. Adamantine. Demantglanz. Diamant. Adamantinus.

    c. Pearly. Perlmutterglanz. Nacre. Margaratinus.

    d. Resinous. Fettglanz. Cire ou gras. Cereus.

    e. Vitreous. Glasglanz. Vitreux. Vitreus.

II. THE ASPECT OF THE FRACTURE. Bruchansehen. Aspect de la cassure. Aspectus internus.

IV. *The Lustre of the Fracture, as in the External Lustre.*

V. *The Fracture.* Der bruch. La cassure ou la surface interieur. Fractura,—*of which are*

  1. *The following varieties,*

    A. *The compact fracture.* Dichte bruch. Dense. Densa.—*This is*

      a. Splintery. Splittrich. Ecailleuse. Festucosa.

      α. Coarse splintery. Grobsplittrich. A grandes ecailles. Festucis majusculis.

β. Small

ß. Small splintery. Kleinsplittrich. A petites ecailles. Festucis minusculis.

γ. Fine splintery. Fein splittrig. A ecailles fines.

b. Even. Eben. Egale ou unie. Æqualis.

c. Conchoidal. Muschlich. Concoide. Conchæformis.

α. With respect to size. Nach der grösse. D'après la grandeur de concavités. Respectu magnitudinis.

    i. *Large conchoidal.* Grossmuschlich. Tres evasé. Grandiuscula.

    ii. *Small conchoidal.* Kleinmuschlich. Peu evasé. Minuscula.

ß. With regard to depth. Nach der tiefe. Profondeur du cavités.

    i. *Deep conchoidal.* Tief muschlich. A cavités profondes.

    ii. *Flat conchoidal.* Flach muschlich. A cavités plates.

γ. With regard to perfection, Nach der auszeichnung. D'aprés la perfection de concavités. Respectu perfectionis.

    i. *Perfect conchoidal.* Volkommen muschlich. Parfait. Perfecta.

    ii. *Imperfect conchoidal.* Unvolkommen muschlich. Imparfait. Imperfecta.

d. Uneven. Uneben. Anguleuse ou inegale. Inæqualis.

α. Coarse grained. Von grobem korne. Grandes inegalités. Granograndi.

ß. Small grained. Von kleinem korne. Petites inegalités. Grano minusculo.

γ. Fine grained. Von feinem korne. Fines inegalités, Grano minuto.

e. Earthy. Erdig. Terreuse. Terrea.

α. Coarse earthy. Groberdig.

ß. Fine earthy. Feinerdig.

f. Hackly. Hakig. Crochu. Hamata.

B. *Split*

B. *Split fracture.* Gespâltener bruch.

(**A.**) *Fibrous fracture* Der fasriche bruch. Fibreuse, Fibrosa. *Here we have to observe,*

a. *The thickness of the fibres.* Die stärke der fasern. Epaisseur des fibres. Crassities fibrarum.

α. *Coarse fibrous.* Grobfasrig. Grosses fibres. Fibris crassiusculis.

β. *Delicate fibrous.* Zartfasrig. Minces fibres. Fibris tenuibus.

b. *The direction of the fibres.* Die richtung der fasern. Formes des fibres. Directio fibrarum.

α. *Straight fibrous.* Geradfasrig. Droites fibres. Fibris rectis.

β. *Curved fibrous.* Krummfasrig. Courbes fibres. Fibris curvis.

c. *The position of the fibres.* Die lage der fasern. Position des fibres. Situs.

α. *Parallel fibrous.* Gleichlaufend fasrig. Fibres paralleles. Fibris parallelis.

β. *Diverging fibrous.* Auseinanderlaufend fasrig. Fibres divergentes. Fibris divergentibus.

i. *Stellular.* Sternförmig. En etoilles. Stellatim.

ii. *Fascicular or scopiform.* Buschelförmig. En faisceaux. Fasciculatim.

γ. *Promiscuous.* Unter oder durcheinanderlaufend fasrig. Fibres croisées. Fibris decussatis.

(**B.**) *The Radiated fracture.* Der strahlich bruch. Rayonnée Radiata. *Here we have to determine*

a. *The breadth of the rays.* Die breite der strahlen. Largeur des rayons. Latitudo radiorum.

α. *Un*

α. Uncommonly broad radiated. Ausserordentlich breitstrahlich. Tres larges. Radiis eximie latis.

4. *Broad radiated.* Breitstrahlich. Larges. Radiis latis.

γ. Narrow radiated. Schmalstrahlich. Etroits. Radiis arctis.

b. The direction of the rays. Die richtung der strahlen. Forme des rayons. Directio.

α. Straight radiated. Geradstrahlich. Droits. Radiis rectis.

β. Curved radiated. Krummstrahlich. Courbes. Radiis curvis,

c. The position of the rays. Die lage der strahlen. Position des rayons. Situs.

α. Parallel. Gleichlaufend. Paralleles. Radiis parallelis.

β. Diverging. Auseinanderlaufend. Divergens. Radiis divergentibus.

i. Stellular. Sternförmig. En entoilles. Stellatim.

ii. Fascicular or scopiform. Buschelförmig. En faisceaux. Fasciculatim.

γ. Promiscuous. Untereinanderlaufend. Croisés ou entrelacés. Radiis decussatis.

d. The passage of the rays, or cleavage. Der durchgang der strahlen, Direction des rayons.

e. The aspect of the rays surface. Das ansehen der strahlichen flächen. Aspect de faces rayonnées.

(c.) *The foliated fracture.* Der blättriche bruch. Feuilletée. Lamellosa.

a. The size of the folia. Die grösse der blätter. Grandeur des feuillets. Magnitudo lamellarum.

b. The

38

b. The degree of perfection of the foliated fracture: Die volkommenheit. Perfection de la cassure feuilletée. Perfectio.

α. *Highly perfect, or specular splendent.* Höchst volkommen oder spiegelflächig blättrich. Tres parfaitement feuilletée. Perfectissime lamellosa.

β. *Perfect foliated.* Volkommen blättrich, Parfaitement feuilletée. Perfectè lamellosa.

γ. *Imperfect foliated.* Unvolkommen blättrich. Imparfaitement feuilletée. Imperfectè lamellosa.

δ. *Concealed foliated.* Vesteckt blättrich. Feuilletée cachée. Confusè lamellosa.

c. The direction of the folia.. Richtung. Forme des feuillets. Directio.

α. *Plane foliated.* Geradblättrich. Droits. Recta.

β. *Curved foliated.* Krummblättrich. Courbes. Curva.

i. *Spherical.* Sphærisch. Spherique. Sphærica.

ii. *Undulating.* Wellenförmig. Ondulé Undulatim.

iii. *Floriform.* Blumig-blättrich. Palmé. Floriformiter.

iv. *Indeterminate.* Unbestimmt. Indeterminé. Indeterminatæ.

d. The position of the folia. Die lage der blätter. Position des feuillets. Situs.

α. *Common foliated.* Gemeinblättrich.

β. *Scaly foliated.* Schuppigblättrich.

e. The aspect of the surface of the folia. Das anschen der blättrichen fläche.

α. *Smooth.* Glatt.

β. *Streaked.* Gestreift.

f. The

f. The passage of the folia or cleavage. Der durchgang der blætter. Clivage ou direction des feuillets. Meatus lamellarum.

α. The number of the cleavages. Zahl der durchgänge.

i. *Single.* Einfach. Simple. Simplex.

ii. *Two-fold.* Zweifach. Double. Duplex.

iii. *Three-fold.* Dreifach. Triple. Triplex.

iv. *Four-fold.* Vierfaeh. Quadruple. Quadruplex.

v. *Six-fold.* Sechsfach. Sextuple. Sextuplex.

β. The angle under which these cleavages intersect one another. Durchsneidungs winkel.

γ. The greater or lesser degree of perfection of each cleavage. Die mehr oder mindere vollkommenheit jedes durchganges.

(D.) *The slaty fracture.* Shiefrige bruch. Cassure schisteuse.

a. Thickness. Stärke. L'epaisseur des feuillets.

a. *Thick slaty.* Dickschiefrig. A feuillets epais.

b. *Thin slaty.* Dunnschiefrige. A feuillets minces.

b. Direction. Richtung. La direction des feuillets.

a. *Straight slaty.* Geradschiefrig. A feuillets plats.

b. *Curved slaty.* Krummschiefrig. A feuillets courbes.

aa. *Indeterminate curved slaty.*

bb. *Undulating curved slaty.*

c. Perfection. Volkommenheit. La perfection.

a. *Perfect slaty.* Volkommen schiefrig. Parfaite.

b. *Imperfect slaty.* Unvolkommen schiefrig. Imparfaite.

d Cleavage. Durchgang. Le sens des feuillets.

aa. *Single.* Einfach. Simple.

bb. *Double.* Zweifach. Double.

2. *Where*

2. *Where several fractures occur at the same time, their relative situation must be observed, as*

A. *One including the other; Fracture in the great.* Bruch im grossem; *and in the small,* Bruch im kleinen.

B. *One traversing the other; Longitudinal and transverse fracture.* Langebruch und queerbruch. *Cross fracture.* Queerbruch.

VI. *The Shape of the Fragments.* Die gestalt der bruchstücke. Forme des fragmens. Figura fragmentorum.

1. *Regular fragments.* Regelmässige bruchsücke. Fragmens reguliers. Fragmenta regularia.

A. *Cubic.* Würfliche. Fr. Cubiques. Fr. Cubica.

B. *Rhomboidal.* Rhomboidalische. Fr. Rhomboidaux. Fr. Rhomboidalia.

   a. Specular on every side. Auf allen seiten spiegelnd. Toutes les faces miroitantes. Omnibus lateribus micantibus.

   b. Specular on four sides. Auf zwei seiten spiegelnd. Deux faces miroitantes. Duobus lateribus micantibus.

C. *Trapezoidal.* Trapezoidische. Fr. Trapzoides. Fr. Trapezoidea.

D. *Three-sided pyramidal, and octahedral.* Dreiseitig pyramidale und octaedrische. Fr. Tetraedres et octaedres. Fr. Pyramidalia et octaedra.

E. *Dodecahedral.* Dodecaedrische. Fr. Dodecaedres. Fr. Dodecaedra.

2. *Irregular*

41

2. *Irregular fragments.* Unregelmässige bruchstücke.
Fr. irreguliers. Fr. irregularia.

A *Cuneiform.* Keilförmige. Fr. cuneiformes. Fr.
cuneiformia.

B. *Splintery.* Splittrige. Fr. esquilleux. Fr. fes-
tucæformia.

C. *Tabular.* Scheibenförmige. Fr. en plaques.
Fr. orbicularia.

D. *Indeterminate angular.* Unbestimmt eckige. Fr.
indeterminés. Fr. indeterminata.

    a. Very sharp-edged. Sehr schärfkantige. A bords
très aigus. Marginibus peracutis.

    b. Sharp-edged. Scharfkantige. A bords aigus.
Marginibus acutis.

    c. Rather blunt-edged. Ein wenig stumpfkantig.
A bords peu aigus. Marginibus parum acutis.

    d. Blunt-edged. Stumpfkantig. A bords obtus.
Marginibus obtusis.

    e. Very blunt-edged. Sehr stumpfkantig. A bords
très obtus. Marginibus perobtusis.

III. THE ASPECT OF THE DISTINCT CONCRETIONS.
Das absonderungs ansehen. Aspect des
pieces separées. Aspectus partium segregatarum.

VII. *The Shape of the Distinct Concretions.* Gestalt der ab-
gesonderten stücke. Forme des pieces separés. Figura partium segregatarum.

   1. *Granular distinct Concretions.* Körnige abgesonderte
stücke. Grenues. Granulosæ : *which differ*

F   A. *In*

A. *In shape.* In der gestalt. Diversité des formes des grains. Figura,—*and this in*

a. *Round granular.* Rundkörnig. Grains arondis. Rotundæ.

α. *Spherical.* Sphærisch. Spheriques. Sphæricæ.

β. *Lenticular.* Linsenförmig. Lenticulaires. Lenticulares.

γ. *Date-shaped.* Dattel-förmig. Dactyliformis.

b. *Angulo-granular.* Eckigkörnig. Anguleux. Angulares.

α. *Common granular.* Gemeinkörnig. Ordinaires. Vulgares.

β. *Longish granular.* Longkörnig. Longues. Longiusculæ.

B. *In magnitude.* In der grösse. Grandeur des pieces separées grenues. Magnitudo.

a. *Large granular.* Grosskörnig. Tres grandes. Grandes.

b. *Coarse granular.* Grobkörnig. Grandes. Majusculæ.

c. *Fine granular.* Feinkörnig. Fines. Minutæ.

2. *Lamellar distinct Concretions.* Schaalig abgesonderte stücke. Lamelleuses ou testacées. Testaceæ,—*which differ*

A. *In the direction of the lamellæ.* Richtung. Diversité des formes des lames. Directio.

a. *Straight lamellar.* Geradschaalig. Planes. Rectæ.

α. *Quite straight.* Ganz gerad. Entierement planes. Perfectè rectæ, *or*

β. *Fortifications-wise bent.* Fortificationsartig gebogen schaalig. En zigzag, Instar munimentorum.

b. *Curved*

b. Curved lamellar. Krummschaalig. Courbes. Cur-
væ.

α. *Indeterminate curved lamellar.* Gemein krummschaalig. Indeterminées. Vulgariter.

β. *Reniform curved lamellar.* Nierförmig gebogen
schaalig. En rognons. Reniformiter.

γ. *Concentrical curved lamellar.* Concentrisch schaalig. Concentriques. Concentricé.

1. *Spherical.* Sphærish. Spheriques. Sphærico-
concentricè.

2. *Conical.* Conisch. Coniques. Conico-con-
centri.

B. *In the thickness.* In der stärke. Epaisseur des lames.
Crassities.

a. *Very thick lamellar.* Sehr dickschaalig. Très epaisses. Crassæ.

b. *Thick lamellar.* Dickschaalig. Epaisses. Crassi-
usculæ.

c. *Thin lamellar.* Dünnschaalig. Minces. Tenues.

d. *Very thin lamellar.* Sehr dünnchaalig Très minces. Tenusssimæ.

3. *Columnar distinct Concretions.* Stänglich abgefonderte
stücke. Colonnaires. Scapiformes : *which are distinguished*

A. *According to the direction.* Nach der richtung. Con-
tournement des colonnes. Directio, *into*

a. *Straight columnar.* Geradstänglich. Droites. Rectæ.

b. *Curved lamellar.* Krummstänglich. Courbés. Cur-
væ.

B. *With regard to thickness.* Stärke. Epaisseur des co
lonnes. Crassities, *into*

a. *Very*

a. Very thick columnar. Sehr dick. Très epaisses et
grandes. Columnares.

b. Thick columnar. Dickstänglich. Epaisses. Crassæ.

c. Thin columnar, or prismatic. Dünnstänglich. Minces. Tenues.

d. Very thin columnar or prismatic. Sehr dünnustäng-
lich. Très minces. Tenuissimæ.

C. *With respect to shape.* Gestalt, *into*

a. Perfect columnar. Volkommen stänglich. Parfaites.
Perfectè.

b. Imperfect columnar. Unvollkommen stänglich. Im-
parfaites. Imperfectè.

c. Cuneiform columnar. Keilförmig stänglich. Cuneiformes. Cuneatim.

d. Ray-shaped columnar. Strahlformig stänglich.

D. *According to the position.* Lage, *into*

a. Parallel. Gleichlaufend.

b. Diverging. Auseinanderlaufend.

c. Promiscuous. Untereinanderlaufend.

4. *In several minerals, two of these varieties, or different
sizes of the same variety of distinct concretions, occur
together, either*

A. *The one including the other, or*

B. *The one traversing the other.*

VIII. *The Surface of the distinct Concretions.* Absonderungs-
fläche. Surface des pieces separées. Superficies partium
segregatarum.

1. *Smooth.* Glatt. Lises. Lævis.

2. *Rough.* Rauh. Rude ou apre. Aspera.

3. *Streaked.* Gestreift. Striée. Striata.

4. *Uneven.* Uneben. Raboteuse. Inæqualis.

IX. *The Lustre of the distinct Concretions.* Absonderungs-
glanz,—*is determined in the same manner as the external
lustre.*

IV. GENERAL

IV. GENERAL ASPECT. Allgemeines ansehen.

X. *The Transparency.* Durchsichtigkeit. Transparence. Pelluciditas.

The degrees are

1. *Transparent.* Durchsichtig. Diaphane. Diaphanum, *either*

 A. *Simply transparent.* Gemein durchsichtig. Diaphane simple. Vulgare.

 B. *Duplicating transparent.* Verdoppelnd durchsichtig. Diaphane double. Duplicans.

2. *Semitransparent* Halbdurchsichtig. Semidiaphane. Semidiaphanum.

3. *Translucent,* Durchscheinend. Transparent. Transparens?

4. *Translucent at the edges.* An den kanten durchscheinend. Transparent aux bords. Marginibus transparens.

5. *Opaque.* Undurchsichtig. Opaque. Opacum.

The Opalescence. Das opalisiren.

 A. *Common or Simple Opalescence.* Das gemeine opalisiren.

 B. *Stellular opalescence.* Das sternförmige opalisiren.

XI. *The Streak.* Der strich. Raclure. Rasura, *is either*

 a. *In regard to colour : it is either*

 a. *Similar to that of the mineral.* Gleich. De même couleur, *or it is*

 b. *Dissimilar.* Verschieden. Du couleur different.

b. In

46

b. In regard to lustre : it remains
  a. *Unchanged.* Unverändert. De même eclat.
  b. *Is increased in intensity.* Nimmt zu. Donnant de l'eclat.
  c. *Is diminished in intensity.* Vermindert. Diminuant de l'eclat.

XII. *The Soiling or Colouring.* Abfarben. Tachure. Tinctura, *by which minerals*
1. *Soil.* Abfärben, *either*
  A. *Strongly.* Stark, *or*
  B. *Slightly.* Etwas, *or*
2. *Do not soil.* Nichtabfärben.
3. *Write.* Schreibend.

V. CHARACTERS FOR THE TOUCH.

XIII. *The Hardness.* Die härte. Dureté. Durities. *The degrees are*
1. *Hard.* Hart. Dur. Durum.
  A. *Resisting the file.* Wird von der feile gar nicht angegriffen. Resistant à la lime. Limæ non cedens.
  B. *Yielding a little to the file.* Wird wenig angegriffen. Cedant un peu à la lime. Limæ parum cedens.
  C. *Yielding to the file.* Wird von der feile stark angegriffen. Cedant à la lime. Limæ cedens.
2. *Semihard.* Halbhart. Demidure. Semidurum.
3. *Soft.* Weich. Tendre. Molle.
4. *Very soft.* Sehr weich. Très tendre. Mollissimum.

XIV. *The*

XIV. *The Tenacity.* Festigkeit. La ductilité. Ductilitas.
The degrees of which are

1. *Brittle.* Spröde. Aigre. Fragile.
2. *Sectile or mild.* Milde. Traitable. Lene.
3. *Ductile.* Geschmeidig. Malleable. Ductile.

XV. *The Frangibility.* Der zusammenhalt. La tenacité. Tenacitas.

1. *Very difficultly frangible.* Sehr schwer zerspringbar. Tres tenace. Tenacissimum.
2. *Difficultly frangible.* Schwer zerspringbar. Tenace. Tenax.
3. *Not particularly difficultly frangible,* or *rather easily frangible.* Nicht sonderlich schwer zerspringbar. Peu tenace. Non multum tenax.
4. *Easily frangible.* Leicht zerspringbar. Cassant facilement. Parum tenax.
5. *Very easily frangible.* Sehr leicht zerspringbar. Cassant très facilement. Valdè parum tenax.

XVI. *The Flexibility.* Die biegsamkeit. Flexibilité. Flexibilitas : *according to which minerals are either*

1. *Flexible.* Biegsam. Flexible, *and this either*
   A. *Elastic flexible.* Elastisch biegsam. Elastique. Elasticè, *or*
   B. *Common flexible.* Gemein biegsam. Ordinaire. Vulgariter, *or*
2. *Inflexible.* Unbiegsam. Inflexible. Inflexibile.

XVII. *The Adhesion to the Tongue.* Dans anhängen an der zunge. Le happement a la langue. Adhæsio ad linguam, *the degrees of which are,*

1. *Strongly adhesive.* Stark an der zunge hängend. Happe beaucoup. Fortiter adhæret.

2. *Pretty*

2. *Pretty strongly.* Ziemlich stark. Assez. Mediocriter.

3. *Weakly, or somewhat.* Etwas. Un peu. Aliquantum.

4. *Very weakly, or a little.* Wenig. Tres peu. Parum.

5. *Not at all.* Gar nicht. Pas du tout. Nihil.

VI. CHARACTERS FOR THE HEARING. Kenzeichen für das gehör.

XVIII. *The Sound.* Der ton. Son. Sonus. *The different sorts of which occurring in the mineral kingdom, are*

1. *A ringing sound.* Klingen. Tintement. Clangor.

2. *A grating sound.* Rauschen. Bruyement. Strepitus. *And*

3. *A creaking sound.* Knirschen. Crissement. Stridor.

II.

IId.

PARTICULAR GENERIC EXTERNAL CHARACTERS OF *FRIABLE*
MINERALS. Besondere generische kennzeichen der zerreiblichen
fossilien.

I. The External Shape. Aussere gestalt. Figure exterieure.
Figura externa. *This is*

1. *Massive.* Derb. Massive. Compactum.
2. *Disseminated.* Eingesprengt. Disseminé. Inspersum.
3. *Thinly coating.* Als dünner überzung. En croute
mince. Superinductum.
4. *Spumous.* Schaumartig. En ecume. Spumæforme; *and*
5. *Dendritic.* Baumförmig. Dendritique. Dendriticum.

II. The Lustre. Glanz. Eclat. Nitor.

1. *The intensity.* Stärke des glanzes. Intensité de l'eclat.
Gradus nitoris.

A. *Glimmering.* Schimmernd. Tremblotant. Micans.
B. *Dull.* Matt. Mat. Nitoris expers.

2. *The sort.* Art des glanzes. Nature de l'eclat. Species
nitoris.

A. *Common glimmering.* Gemein schimmernd. Ordinaire. Vulgaris.
B. *Metallic glimmering,* Metallischschimmernd. Metallique. Metallicus.

G   III. The

III. The Aspect of the Particles. Ansehen der theilchen. L'aspect des parties. Aspectus particularum.

1. *Dusty.* Staubige. Pulverulentes. Pulveriformes.

2. *Scaly.* Schuppige. Ecailleuses. Squamosæ.

IV. The Soiling or Colouring. Abfärben. La tachure. Tinctura.

1. *Strongly.* Stark. Beaucoup. Multum tingens.

2. *Slightly.* Wenig. Peu. Parum.

V. The Friability. Zerreiblichkeit. Friabilité. Friabilitas,

1. *Loose.* Lose. Incoherant. Non conglutinatæ.

2. *Cohering.* Zusammengebacken. Coherant, Conglutinatæ.

IIId.

IIId.

PARTICULAR GENERIC EXTERNAL CHARACTERS OF *FLUID* MINERALS. Besondere generische Kennzeichen der flüssigen fosilien.

I. The Lustre.   Glanz.   Eclat.   Nitor.
   1. *Metallic.*   Metallischer.   Metallique.   Metallicus.
   2. *Common.*   Gemeiner.   Ordinaire.   Vulgaris.
II. The Transparency.   Durchsichtigkeit.   Transparence. Pelluciditas.
   1. *Transparent.*   Durchsichtig.   Diaphane.   Diaphanum.
   2. *Troubled, or turbid.*   Trübe.   Troublé.   Turbidum.
   3. *Opaque.*   Undurchsichtig.   Opacum.
III. The Fluidity.   Flüssigkeit.   Fluidité.   Fluiditas.
   1. *Fluid.*   Flüssig.   Parfaite.   Fluidum.
   2. *Viscid.*   Zahe.   Viscuse.   Lentum.

REMAINING GENERAL GENERIC EXTERNAL CHARACTERS. Uebrige allgemeine generische aussere Kennzeichen.

III. The Unctuosity.   Fettigkeit.   Toucher ou gras.   Pinguitudo.

Of this we have the following degrees.

   1. *Very greasy,*   Sehr fett.   Fort gras.   Pinguissimum.
   2. *Greasy*   Fett.   Gras.   Pingue.
   3. *Rather greasy.*   Ein wenig fett.   Un peu gras.   Parum pingue.
   4. *Meagre.*   Mager.   Maigre.   Macrum.

V. The

IV. The Coldness. Kälte. Froid. Frigus.

With respect to which minerals are

1. *Very cold.* Sehr kalt. Tres froid.
2. *Cold.* Kalt. Froid. Frigidum.
3. *Pretty cold.* Ziemlich kalt. Mediocriment froid. Frigidiusculum.
4. *Rather cold.* Wenig kalt. Mediocrement fröid. Parum frigidum.

V. The Weight. Schwere. La pesanteur specifique. Gravitas.

1. *Swimming or supernatant.* Schwimménd. Surnageant. Natans.
2. *Light.* Leichte. Leger. Levis.
3. *Not particnlarly heavy, or rather heavy.* Nicht sonderlich schwer. Mediocrement pesant. Parum gravis.
4. *Heavy.* Schwer. Pesant. Gravis.
5. *Uncommonly heavy.* Ausserordentlich schwer. Tres pesant. Eximiè gravis.

VI. The Smell. Geruch. Odeur. Odor.

1. *Spontaneously emitted.* Für sich.

   A. *Bituminous.* Bituminös. Bitumineuse. Bituminosus.
   B. *Faintly sulphureous.* Schwach schweflich. Legerement sulphureuse. Sulphureus.
   C. *Faintly bitter.* Schwach bitterlich. Legerement amer. Subamarus.

2. *Produced by breathing on it.* Nach dem anhauchen. En y portant la vapeur de l'expiration. Adflatu.

   A. *Clay-like smell.* Thonigen geruch. Argilleuse. Argillosus.

3. *Excited by friction.* Durch reibung. Par la frottement. Frictione.

   A. *Urinous.*

A. *Urinous.* Urinös. Urineuse. Urinosus.

B. *Sulphureous.* Schweflich. Sulphureuse. Sulphu-
ratus.

C. *Garlick-like, or arsenical.* Knoblauchartig. Ail.
Alliaceus.

D. *Empyreumatic.* Empyreumatisch. Empyreume.
Empyreumaticus.

VII. THE TASTE. Geschmack. Saveur. Sapor.

The varieties are

1. *Sweetish taste.* Süssalzig. Salée. Dulce salsus.
2. *Sweetish astringent.* Sulszusamenziehend. Adstrin-
gente.
3. *Styptic.* Herbe. Acerbe. Stypticus.
4. *Saltly bitter.* Salzigbitter. Salée amere. Salso-
amarus.
5. *Saltly cooling.* Salzigkühlend. Salée fraiche. Fri-
gido-salsus.
6. *Alkaline.* Laugenhaft. Alcaline. Lixiviosus.
7. *Urinous.* Urinös. Urineuse. Urinosus.

EXTERNAL

EXTERNAL CHARACTERS

OF

MINERALS.

I.

COLOUR.

WE begin our description of the External Characters
of Minerals with that of Colour, as it is the character
which first particularly strikes the eye. It exhibits very
great variety, and hence its determination is often attended
with considerable difficulty. Although it is an important
and useful character, it was but ill understood before the
time of WERNER, and it is even at present, by some mineralogists, considered as of little or no value. The older
mineralogists had no very accurate Nomenclature of Colours, and rarely gave any definition of them; hence it
was, that this character, in their systems, did not afford
satisfactory descriptions. Some modern mineralogists,
particularly

particularly those of the French School, use in their de‑
scriptions, only single, and often unconnected varieties
of colour, which is an erroneous practice ; because in
describing species, we ought to enumerate all the varie‑
ties they exhibit, and in a natural order, so that we may
obtain a distinct conception of the arrangement of these
varieties into groups or suites that characterise the spe‑
cies. WERNER was early aware of the utility of this
character, and, by a careful study of all its appearances
and varieties. was enabled to form a system of colours for
the discrimination of minerals, in which he established a
certain number of fixed or standard colours, to which all
the others could be referred, defined the varieties and ar‑
ranged them according to their resemblance to these
standard colours, and placed them in such manner, that
the whole colours in the system formed a connected se‑
ries.

In establishing the fixed or standard colours, he thought
he could not do better than adopt those as simple co‑
lours, which are considered as such in common life ; of
these he enumerates eight, which he denominates *chief or
principal* colours ; they are *white, grey, black, blue, green,
yellow, red,* and *brown.* Although several of these colours
are physically compound, yet for the purposes of the
oryctognost it is convenient to consider them as simple.

WERNER remarks, " I could not here enter into an a‑
" doption of the seven colours into which the solar ray is
" divided by the prism, as principal colours, nor into the
" distinction of the colours accordingly as they are either
" simple or compound ; nor could I omit white and black,
" the former being considered as a combination of all co‑
" lours, and the latter as the mere privation of light or
" colour ; for these are distinctions that pertain to the
" theory

" theory of colours among natural philosophers, and can-
" not be well applied in common life, in which black is
" ranked among the colours as well as white and yellow ;
" and green, which is mixed, considered as a principal
" colour, as well as red, which is simple.

" In the adoption of the principal colours enumerated
" above, I am countenanced by Dr Schœffer, who has
" exhibited them with the exception of the grey, in his
" sketch of a general association of colour, Regensburg,
" 1769. I am, however, justified in adding the grey
" colour, by observing, that it occurs very frequently in
" the mineral kingdom ; that the attempt to bring it un-
" der any one of the other colours would be attended
" with many difficulties, and that, if we have respect to
" denominations, it is considered in common life as ac-
" tually differing from the others."—Werner's *External
Characters*, p. 38, 39.

Each of these principal colours contains one which is,
oryctognostically considered, pure or unmixed with any
other, which is called the *characteristic colour :* thus snow-
white is the characteristic colour of white ; ash-grey, of
grey ; velvet-black, of black ; Berlin-blue, of blue ; emerald-green, of green ; lemon-yellow, of yellow ; carmine-
red, of red ; and chesnut-brown, of brown.

Having thus established eight characteristic colours, he
next defined and arranged the most striking subordinate
varieties.

The definitions were obtained principally by ocular examination, which enables us speedily to detect the different colours of which the varieties are composed. In de-
tailing the results of this kind of *ocular analysis*, if I may
use the expression, the predominant component parts are
mentioned first, and the others in the order of their quan-

H   tity.

tity. Thus apple-green is found to be a compound colour, and we discover by comparing it with emerald-green, that it is principally composed of that colour and another, which is greyish white; we therefore define apple-green to be a colour composed of emerald green and a small portion of greyish-white. The method he followed in arranging the varieties is simple and elegant. He placed together all those varieties which contained the same principal colours in a preponderating quantity, and he arranged them in such a manner, that the transition of the one variety into the other, and of the principal colour into the neighbouring ones, was preserved. To illustrate this by an example. Suppose we have a variety of colour which we wish to refer to its characteristic colour, and also to the variety under which it should be arranged. We first compare it with the principal colours, to discover to which of them it belongs, which in this instance we find to be green. The next step is to discover to which of the varieties of green in the system it can be referred. If, on comparing it with emerald green, it appears to the eye to be mixed with another colour, we must, by comparison, endeavour to discover what this colour is; if it prove to be *greyish-white*, we immediately refer the variety to *apple-green*; if, in place of *greyish-white*, it is intermixed with *lemon-yellow*, we must consider it *grass-green*; but if it contains neither greyish-white nor lemon-yellow, but a considerable portion of *black*, it forms *blackish-green*. Thus, by mere ocular inspection, any person accustomed to discriminate colours correctly, can ascertain and analyse the different varieties of colour that occur in the mineral kingdom.

The transition of the principal colours and their varieties into each other, he represents by placing the characteristic

teristic colours in the middle of a series of which all the members are connected together by transition, and whose extreme links connect them with the preceding and following principal colours. Thus, emerald-green is placed in the middle of a series, the members of which pass, on the one hand, by increase of the proportion of blue into the next colour-suite, the blue; on the other hand, by the increase of yellow into yellow, siskin green forming the connecting link with yellow, and verdigris green with blue.

NAMES OF THE COLOURS.

The names of the colours are derived, 1st, From certain bodies in which they most commonly occur, as milk-white, siskin-green, liver-brown; 2d, From metallic substances, as silver-white, iron-black, and gold-yellow; 3d, From names used by painters, as indigo blue, verdigris-green, and azure-blue; 4th, From that colour in the composition which is next in quantity to the principal colour, as bluish-grey, yellowish-brown, &c.; and, 5th, From the names of persons, as Isabella-yellow, now called Cream-yellow.

The principal colours are divided into two series, the one comprehending what WERNER terms *bright colours*, the other *dead colours;* red, green, blue, and yellow belong to the first; and white, grey, black, and brown, to the second.

ARRANGEMENT OF THE COLOURS.

The different characteristic colours and their varieties pass into each other, forming suites of greater or less extent, in which the colours either differ more and more

from

from the first member of the series, as they approach the
extremity, thus forming *straight series*, or, after reaching
a certain point of greatest difference from the first colour,
again gradually approach, and at length pass into it ; thus
forming *circular series*. In this way the eight principal
colours pass into each other in the order in which we
have already enumerated them, and thus form a straight
series. The blue colour, however, after it has passed
through green and yellow into red, passes from this
latter colour by several intermediate varieties again into
blue, thus forming a circular series or group.

In the system of colours, we do not introduce these various subordinate transitions and series, but simply arrange all the colours as they pass into each other, beginning with the white, and ending with the brown. The
varieties of most of the different principal colours are so
arranged, that their characteristic colour is placed in the
middle of the series, and all those varieties that incline to
the preceding principal colour, are placed immediately
after it ; while those that incline to the next or following principal colour immediately precede it. This,
however, is not the case with the white and grey colours ;
therefore the characteristic colours in those series do not
stand in the middle ; on the contrary, in the white, it is
placed at the beginning, and in the grey at the end.

DEFINITIONS

I. DEFINITIONS OF THE DIFFERENT VARIE-
TIES OF COLOUR.

A. WHITE.

This is the lightest of all the colours ; hence the slight-
est intermixture of other colours becomes perceptible. The white colour occurs principally in
earthy and saline minerals, seldom in metalliferous minerals, and very rarely amongst inflammable minerals. The following are the varieties of this colour.

a. *Snow-white* is the purest white colour, being free
of all intermixture, and is the only colour of this
suite which has no grey mixed with it. It resembles new-fallen snow. As example of it, we
may mention Carrara marble.

b. *Reddish-white* is composed of snow-white, with a
very minute portion of crimson-red and ash-
grey. It passes into flesh-red. Examples, porcelain earth and rose-quartz.

c. *Yellowish-white* is composed of snow-white, with
very little lemon-yellow and ash-grey. It passes
on the one side into yellowish-grey, on the other
into straw-yellow. Examples, chalk, limestone
and semiopal.

d. *Silver-white* is the colour of native silver, and is
distinguished from the preceding by its metallic
lustre. Examples, arsenical pyrites and native
silver.

e. *Greyish-white* is snow-white mixed with a little ash-
grey. Examples, quartz and limestone.

f. *Greenish-white* is snow-white mixed with a very
little emerald-green and ash-grey. It passes
into apple-green. Examples, amianthus, foliated
limestone, and amethyst.

g. *Milk-*

   g. Milk-white is snow-white mixed with a little Berlin-blue and ash-grey. It passes into smalt-blue. The colour of skimmed milk. Examples, calcedony and common opal.

   h. Tin-white differs from the preceding colour principally in containing a little more grey, and having the metallic lustre. It passes into pale lead-grey. Examples, native antimony and native mercury.

B. GREY.

This, which is one of the palest colours, is a compound of white and black, so that it forms the link by which these two colours are connected together, and is therefore placed between them. It occurs very frequently in the mineral kingdom. The following are its varieties.

   a. Lead-grey is composed of light ash-grey with a small portion of blue, and possesses metallic lustre. It contains the following subordinate varieties.

    α. Whitish lead-grey. It is a very light lead-grey colour, into the composition of which a considerable portion of white enters, and therefore nearly approaches to tin-white. Example, native arsenic on the fresh fracture.

    β. Common lead-grey. It is the purest lead-grey, with a slight intermixture of yellow. Example, common grey antimony-ore.

    γ. Fresh lead grey. It contains rather more blue than the preceding variety, with a slight tint of red, so that it has what is called a fresh or burning aspect. Examples, galena or lead-glance, and molybdena.

                δ. Blackish

з. Blackish lead-grey. Is common lead-grey mixed with a little black. Examples, silver-glance
or sulphureted silver, and copper-glance or vitreous copper-ore.

b. Bluish-grey is ash-grey mixed with a little blue, or
is lead-grey without metallic lustre. Examples,
hornstone and limestone.

c. Pearl-grey is pale bluish-grey intermixed with a
little red. It passes into lavender-blue. Examples, quartz, porcelain jasper, crystallised hornstone, and a very pale variety of pearl.

d. Smoke-grey is dark bluish-grey, mixed with a little
brown. Examples, flint, and some varieties of
fluor-spar.

e. Greenish-grey is ash-grey mixed with a little emerald-green, and has sometimes a faint trace of
yellow. It passes into mountain-green Examples, clay-slate, whet-slate, potstone, sometimes mica, prehnite, and cat's-eye.

f. Yellowish-grey is ash-grey mixed with lemon-yellow and a minute trace of brown. It sometimes
passes into cream-yellow and wood-brown. Examples, calcedony and mica.

g. Ash-grey is the characteristic colour. It is a compound of yellowish-white and brownish-black. It
is the colour of wood-ashes. It passes on the
one hand into greyish-black, on the other into
greyish-white, as also into greenish, greyish, and
smoke-grey. It seldom occurs pure in the mineral kingdom. Examples, quartz, flint and mica.

h. Steel-grey is dark ash-grey with metallic lustre.
It is the colour of newly broken steel. Examples, grey copper-ore and native platina.

C. BLACK.

C. BLACK.

It presents fewer varieties than any of the other colours, owing probably to the intermixture of
lighter colours not being observable in it. The
discrimination of its varieties is attended with
considerable difficulty, and can only be satisfac-
torily accomplished after much practice. The
following are its varieties.

a. *Greyish-black* is velvet-black mixed with ash-grey.
It passes into ash-grey. Is very distinct in basalt.

b. *Iron-black* is principally distinguished from the preceding variety by its being rather darker, and
possessing a metallic lustre. It passes into steel-
grey. Examples, magnetic ironstone and iron-
mica.

c. *Velvet-black* is the characteristic colour of this series. It is the colour of black velvet. Example,
obsidian.

d. *Pitch-black*, or *brownish-black*, is velvet-black mixed with a little yellowish-brown. It passes into
blackish-brown. Example, earthy cobalt-ochre
and mica.

e. *Greenish black*, or *raven-black*, is velvet-black mixed with a little greenish-grey. It passes into
blackish-green. Example, hornblende.

f. *Bluish-black* is velvet-black mixed with a little
blue. It passes into blackish blue, and appears
sometimes to contain a slight trace of red. Example, black earthy cobalt-ochre.

D. BLUE.

D. BLUE.

The characteristic colour, which is Berlin-blue, is placed in the middle of the series, and all those varieties that contain red in their composition, on the one side, and those containing green, on the other. It is rarer among minerals than the preceding ; blackish blue connects it with black, sky blue with green ; and it is connected with red by violet-blue and azure blue. The following are its varieties.

a. *Blackish-blue* is Berlin-blue mixed with much black and a trace of red. It passes, on the one side, into bluish black, on the other, into azure-blue. Example, azure copper-ore.

b. *Azure-blue* is Berlin blue mixed with a little red. It is a burning colour. Examples, azure copper-ore, and azure-stone.

c. *Violet-blue* is Berlin blue mixed with much red and very little black. It borders on columbine-red. It is the tint of colour we observe in the violet when it is about to blow. It is the most frequent of the blue colours. Examples, amethyst and fluor-spar.

d. *Lavender-blue* is violet-blue, intermixed with a small portion of grey. It is intermediate between pearl-grey and violet-blue. Examples, lithomarge and porcelain-jasper.

r. *Plum-blue* is Berlin blue, with more red than in violet-blue, and a small portion of brown and black. It passes into cherry-red and broccoli-brown. Example, spinel.

I   *f. Berlin-blue*

f. Berlin-blue is the purest or characteristic colour of the series. Example, sapphire.

g. Smalt-blue is Berlin-blue with much white, and a trace of green. It passes into milk-white. It occurs in pale-coloured smalt, named Eschel, and also in blue iron-earth, and earthy azure copper-ore.

h. Duck-blue is a dark blue colour, composed of blue, much green, and a little black. Examples, frequently in ceylanite, and in a rare variety of indurated talc.

i. Indigo-blue, a deep blue colour, composed of blue, with a considerable portion of black and a little green. Example, blue iron-earth of Eckardsberg in Thuringia.

k. Sky-blue is a pale blue colour, composed of blue and green, and a little white. It forms the link which connects the blue series with the green. It is named Mountain-blue by painters. It is the colour of a clear sky, and hence its name. It occurs but rarely in the mineral kingdom. Example, lenticular ore.

E. GREEN.

The varieties of this colour naturally fall into two principal suites; in the one of which the blue colour prevails; in the other the yellow; and between the two is placed the pure or characteristic colour, the emerald-green. Although it is not a common colour in the mineral kingdom, yet it is met with more frequently than the blue. In earthy minerals the green colours are generally

rally owing to oxide of iron ; and in a few
cases to the oxide of chrome ; and in very few
to oxide of nickel. Green colours also occur in
several of the ores of copper.

The following are the varieties of this colour.

a. *Verdigris-green* is emerald-green mixed with much
Berlin-blue, and a little white. It is the link
which connects the green and blue colours together. Examples, copper green and Siberian
felspar.

b. *Celandine green* is verdigris-green mixed with ash-
grey. Examples, green-earth, Siberian and Brazilian beryl.

c. *Mountain-green* is composed of emerald green, with
much blue, and a little yellowish-grey ; or verdigris-green with yellowish-grey. It passes into greenish-grey. Examples, beryl, aqua ma-
rine topaz, glassy actynolite, common garnet,
and hornstone.

d. *Leek-green* is emerald-green, with bluish-grey and
a little brown. It is the Sap-green of painters.
In this colour, the blue and yellow colours are
in equal proportion, so that neither preponde-
rate. Examples, nephrite, common actynolite,
and prase.

e. *Emerald-green.* The characteristic or pure unmixed green. All the preceding green colours are
more or less mixed with blue, and at length pass
into it ; but the following part of the green series, by the increasing proportion of yellow, at
length passes into yellow. Examples, emerald,
fibrous

fibrous malachite, copper-mica, and sometimes also fluor-spar.

f. *Apple green* is emerald green mixed with a little greyish white. It passes into greenish white. Examples, nickel-ochre and chrysoprase.

g. *Grass green* is emerald green mixed with a little lemon yellow. The colour of fresh newly sprung grass. Example, uranite.

h. *Blackish-green* is pistachio-green mixed with a considerable portion of black. It passes into greenish-black. Examples, precious serpentine and augite.

i. *Pistachio-green* is emerald green mixed with more yellow than in grass-green, and a small portion of brown. Examples, chrysolite, and epidote or pistacite.

k. *Asparagus green* is pistachio-green mixed with a little greyish-white; or emerald-green mixed with yellow and a little brown. It passes into liver-brown. Examples, garnet, oliven-ore and beryl.

l. *Olive-green* is grass-green mix d with much brown and a little grey. It passes into liver-brown. Examples, common garnet, oliven ore, pitch-stone, and epidote or pistacite.

m. *Oil-green* is emerald green mixed with yellow, brown and grey; or pistachio-green, with much yellow and light ash-grey. It is the colour of fresh vegetable oil. Examples, fuller's-earth, beryl and pitch-stone.

n. *Siskin-green* is emerald-green mixed with much lemon-yellow and a little white. It makes the transition

transition to the yellow colour. Examples, uran-
mica and steatite.

F. YELLOW.

Among the varieties of this species of colour, there
are three possessing metallic lustre, viz. brass-
yellow, gold yellow, and bronze-yellow. The
characteristic colour is lemon-yellow ; which is
placed in the middle of the series ; the colours
which precede it are greenish yellow and those
which follow it are reddish yellow. The one
side of the series, by the increase of the green,
passes by sulphur-yellow into green ; the other,
by the increase of red, passes, by means of o-
range-yellow, into red. It is a frequent colour
in the mineral kingdom. The following are its
varieties.

a. *Sulphur-yellow* is lemon-yellow mixed with much
emerald green and white. It is the colour of native sulphur. Example, native sulphur.

b. *Brass-yellow* differs from the preceding colour principally in having a metallic lustre ; it contains
a small portion of grey. Example, copper-pyrites.

c. *Straw-yellow* is sulphur-yellow mixed with much
greyish-white. It passes into yellowish-white
and yellowish grey. Examples, calamine, serpentine, and yellow cobalt-ochre.

d. *Bronze-yellow* is brass-yellow mixed with a little
steel-grey, and a minute portion of reddish brown.
The colour of bell-metal. Example, iron-pyrites.

e. *Wax-yellow*

e. *Wax-yellow* is composed of lemon-yellow, reddish-
brown, and a little ash-grey; or it may be considered as honey-yellow with greyish white. It
is the colour of pure unbleached wax. Examples, opal and yellow lead-ore.

f. *Honey-yellow* is sulphur-yellow mixed with chesnut brown. It passes into yellowish brown.
Examples, fluor-spar and beryl.

g. *Lemon-yellow* is the pure unmixed colour. It is the
colour of ripe lemons. Example, yellow orpiment.

h. *Gold-yellow* is the preceding colour with metallic
lustre. Example, native gold.

i. *Ochre yellow* is lemon-yellow mixed with a considerable quantity of light chesnut brown. It
passes into yellowish brown. It is a very common colour among minerals. Examples, yellow
earth and jasper.

k. *Wine-yellow* is lemon-yellow mixed with a small
portion of red and greyish-white. The colour of
Saxon home-made wine. Examples, Saxon and
Brazilian topaz.

l. *Cream-yellow* or *Isabella-yellow*. It contains more
red and grey than the wine-yellow, and also a
little brown. It passes into flesh-red. Examples, bole from Strigau, and compact limestone.

m. *Orange-yellow* is lemon-yellow with carmine-red.
It is the colour of the ripe orange. Examples,
streak of red orpiment, and also uran-ochre.

G. RED.

G. RED.

It exhibits more varieties than the other colours, and
is very common in the mineral kingdom. The
characteristic colour is carmine-red ; all the o-
thers incline either to yellow or blue ; hence
there are two principal suites ; the first of which
contains yellowish-red colours ; the second bluish-
red colours. The red colours are principally
owing to oxides of iron, manganese and cobalt,
and combinations of metals with sulphur and arsenic. The following are the varieties.

a. *Aurora* or *morning red* is carmine-red mixed with
much lemon-yellow. It passes into orange-yellow. Example, red orpiment.

b. *Hyacinth-red* is carmine-red mixed with lemon-
yellow and a minute portion of brown ; or aurora-red mixed with a minute portion of brown.
It passes into brown. Examples, hyacinth and
tile-ore.

c. *Tile-red* is hyacinth-red mixed with greyish-white.
It is the colour of tiles or bricks. Examples,
porcelain-jasper and zeolite.

d. *Scarlet-red* is carmine-red, mixed with a very little
lemon-yellow. It is a well-known colour of
much intensity. Example, light-red cinnabar
from Wolfstein.

e. *Blood-red* is scarlet-red mixed with a small portion
of black. Examples, pyrope and jasper.

f. *Flesh-red* is blood-red mixed with greyish-white.
Examples, felspar, calcareous-spar, and straight
lamellar-heavy spar.

g. *Copper-red.* It scarcely differs from the preceding

ing variety, but in possessing a metallic lustre.
Examples, native copper and copper-nickel.

h. *Carmine-red* is the characteristic colour. Example,
spinel, articularly in thin splinters.

i. *Cochineal-red* is carmine-red mixed with bluish-
grey. Examples, dark-red cinnabar and red copper-ore.

k. *Crimson-red* is carmine-red mixed with a considerable portion of blue. Example, oriental ruby.

l. *Columbine-red* is carmine-red, with more blue than
the preceding variety, and, what is characteristic
for this colour, a little black. Example, oriental garnet.

m. *Rose-red* is cochineal-red mixed with white. It
passes into reddish-white. Examples, red manganese-ore and quartz.

n. *Peach blossom-red* is crimson-red mixed with white.
Example, red cobalt ochre.

o. *Cherry-red* is crimson-red mixed with a considerable portion of brownish-black. Examples, spinel, red antimony-ore, and precious garnet.

p. *Brownish-red* is blood-red mixed with brown. It
passes into brown. Example, clay-ironstone.

H. BROWN.

This, after black, is the darkest colour in the system.
The whole species or suite can be distinguished
into those which have red, and those which
have yellow mixed ; between these is placed the
fundamental colour, the pure unmixed chesnut-brown, and the last variety, from the
quantity of black it contains, connects the
brown series with the black. Varieties of this
colour

colour occur frequently in the mineral kingdom,
particularly among the ores of iron, and the inflammable minerals.

a. *Reddish-brown* is chesnut-brown mixed with a little
red and yellow ; or chesnut-brown with a small
portion of aurora-red. It passes into brownish-
red. Example, brown blende from the Hartz.

b. *Clove-brown* is chesnut brown, mixed cochineal-
red, and a little black. It is the colour of the
clove. It passes into plum blue and cherry-
red. Examples, rock-crystal, brown hematite,
and axinite.

c. *Hair-brown* is clove-brown mixed with ash-grey.
Example, Cornish tin-ore.

d. *Broccoli-brown* is chesnut-brown mixed with much
blue, and a small portion of green and red. It
passes into cherry-red and plum-blue. It is a
rare colour. Example, zircon.

e. *Chesnut-brown.* Pure brown colour. It is a rare
colour. Example, jasper.

f. *Yellowish-brown* is chesnut brown mixed with a
considerable portion of lemon yellow. It passes
into ochre-yellow. It is one of the most common colours in the mineral kingdom. Examples,
iron-flint and jasper.

g. *Pinchbeck-brown* is yellowish-brown with metallic
lustre. Rather the colour of tarnished pinchbeck. Example, mica.

h. *Wood-brown* is yellowish-brown mixed with much
pale ash-grey. It passes into yellowish-grey.
Examples, mountain wood, and bituminous
wood.

K   *i. Liver-*

> *i. Liver-brown* is chesnut-brown mixed with olive-
> green and ash-grey. It is the colour of boiled,
> not fresh liver. It passes into olive-green. Ex-
> ample. common jasper.
>
> *k. Blackish-brown* is chesnut-brown mixed with black.
> It passes into brownish black. Examples, mine-
> ral pitch from Neufchatel, moor-coal, and bitu-
> minous wood.

The immense variety of colours that occur in the mineral kingdom, constitute an almost infinite series, to
characterise every individual of which is next to impos-
sible. The colours we have already defined, are a few
only of the most prominent features of that great and
beautiful series, and serve as points of comparison, and
as the boundaries between which every occurring colour
lies.

From the small number of colours we have defined,
and the great variety that occur in minerals, it is evident
that the greater number of occurring colours will not correspond exactly with those defined, but will lie between
them. It is this circumstance in particular that renders
it so difficult to get an acquaintance with colours. To
obviate this in some degree, WEBNER uses terms which
express correctly certain prominent differences which are
to be observed between every two colours. Thus when
one colour approaches slightly to another, it is said to
incline towards it, (es nährt sich) ; when it stands in the
middle between two colours, it is said to be *intermediate*,
(es steht in der mitte) ; when, on the contrary, it evidently approaches very near to one of the colours, it is
said to *fall* or *pass* into it, (es geht über).

II.

II. THE INTENSITY OR SHADE OF THE COLOURS.

Each colour can be distinguished according to its relative *intensity*, of which, as expressed in the tabular view,
there are four degrees, viz. *dark, deep, light,* and *pale.*
Thus the principal colours can be divided into four classes, according to their degrees of intensity ; blue, black,
and brown are dark ; green and red are deep ; yellow and
white are light ; and grey is pale. But this distinction,
as far as regards the principal colours, may be dispensed
with, as they have been already sufficiently discriminated
by their division into bright and dead ; it may therefore
be confined to the varieties. Thus, in the blue series,
lavender-blue is pale ; smalt and skye blue, light ; Berlin,
azure, and violet-blue, deep ; and indigo, plum, and
blackish-blue are dark colours.

The intensity of the colour of a fossil depends often on
its degree of transparency ; for the more transparent it is,
its colour is the paler ; and the more opaque, the colour
is the darker. Many transparent minerals have therefore a very pale colour, which has caused some mineralogists to describe them as colourless, which, however, is
not the case, as their shade of colour is easily detected
by an experienced eye, and it can even be discovered, by
comparing the minerals with another, by those who have
been little accustomed to such investigations.

The intensity of the lustre has also a considerable effect on the intensity of the colour.

III.

III. THE DELINEATIONS OR PATTERNS FORMED BY THE COLOURS.

The distinctions included under this head depend on the shape which the colour assumes. It is only to be observed on simple minerals; therefore, those mineralogists who have attempted to consider it as a character for compound minerals, have deceived themselves. It belongs in general to the individual. The following are the differrent kinds enumerated and described by WERNER.

A. *Dotted.* In this variety dots or small spots are irregularly dispersed over a surface which has a different colour from the spots. It occurs frequently in serpentine, but seldom in other minerals.

B. *Spotted.* If the spots are from a quarter of an inch to an inch in diameter, and the basis or ground still visible, it is said to be spotted. It is either *round* and *regul rly spotted,* or *irregularly spotted.* The first occurs in clay-slate; the second in marble.

C. *Clouded.* Here no basis is to be observed; the boundaries of the colours are not sharply marked, and the spots run into each other. It occurs in marble and jasper.

D. *Flamed.* When the spots are long and acuminated, and arranged according to their length, the flamed delineation is formed. It has still a basis. It occurs in striped jasper, marble, &c.

E. *Striped.*

E. *Striped.* Consists of long and generally parallel stripes that touch each other and fill up the whole mass of the stone, so that it has no ground. It presents two varieties.

    a. *Straight striped,* as in striped jasper and variegated clay.

    b. *Ring-shaped,* occurs in Egyptian jasper.

F. *Veined.* Consists of a number of more or less delicate veins crossing each other in different directions, so that it is sometimes net-like. We can always distinguish a base or ground. Examples, black marble veined with calcareous spar or quartz, jasper and serpentine.

G. *Dendritic.* Represents a stem with branches, on a ground. Examples, steatite and dendritic calcedony.

H. *Ruiniform.* Resembles ruins of buildings. It occurs in Florentine marble, which is from this circumstance called *landscape marble.*

These colour-delineations occur most frequently in marble, jasper, and serpentine, and are characteristic of them. They occur seldom in gypsum, flint, calcedony, &.c.

IV.

IV. THE PLAY OF THE COLOURS.

If we look on a mineral which possesses this proper.
ty, we observe. on turning it slowly, besides its common
colours, many others, which are bright, change very
rapidly, and are distributed in small spots or patches.
A strong light is required, in order to see this appearance
distinctly, and it never occurs in opaque or feebly translucent minerals. We observe it in the diamond when
cut, in precious opal, and in the fire marble of Bleyberg
in Carinthia. It appears to the greatest advantage in
sunshine, probably, however, even more beautiful in
candle light.

V. THE CHANGEABILITY OF THE COLOURS.

When the surface of a mineral, which is turned in different directions, exhibits, besides its common colours,
different bright colours, which do not change so rapidly,
are fewer in number, and occur in larger patches than in
the play of the colour, it is said to exhibit what is called
the changeability of the colours.

We distinguish two kinds of this phenomenon.

A. That which is observed by looking in different
positions *on* the mineral, as in Labradore felspar.

B That observed by looking *through* it, as in the
common opal, which shews a milk-white colour
when we look on its surface, but when held between the eye snd the light is wine-yellow.

VI. THE

VI. THE IRIDESCENCE.

When a mineral exhibits the colours of the prism or the rainbow, arranged in parallel, and sometimes variously curved layers, it is said to be iridescent. It is to be observed by

> *A.* Looking *on* the mineral, as in the variety of calcareous spar, called Iceland or duplicating spar, adularia, beryl, &c. and by
>
> *B.* Looking *through* it, as in rainbow calcedony.

VII. TARNISHED COLOURS.

That this character is of importance, is evident from its frequent occurrence among minerals, and the attention which WERNER has dedicated to its developement.

A mineral is said to be tarnished, when it shews on its external surface, or on that of the distinct concretions, fixed colours different from those in its interior or fresh fracture.

These colours are distinguished, according to their origin ; some minerals shewing them,

> *a.* In the bosom of the earth, as specular iron-ore or iron-glance and radiated grey antimony-ore : others
>
> *b.* On the exposure of the recent fracture to the action of the air, as variegated copper-ore, and native arsenic.

In

In those minerals where the tarnish has taken place in the bosom of the earth, no new tarnish takes place on exposure of a fresh fracture to the air.

They are further divided according to their kind. Minerals in whatever manner they receive their tarnish are, *a. Simple*, or *b. Variegated.* When we say a colour is simple, we mean that one colour predominates over the whole surface; variegated, when the surface shews many different colours which are distinct or run into each other.

Of the merely simple tarnished colours we may mention as examples the following :

   α. Grey,—white cobalt-ore.

   β. Black,—native arsenic.

   γ. Brown,—magnetic pyrites.

   δ. Reddish,—native bismuth.

The variegated or party-coloured, are distinguished according to the intensity of their basis. Of these the following are enumerated in the tabular view.

   α. Pavonine, or *Peacock-tail tarnish.* This is an assemblage of yellow, green, blue, red, and brown colours, on a yellow ground. The colours are nearly equal in proportion, and are never precisely distinct, but always pass more or less into one another. Example, copper pyrites.

   β. Iridescent, or *Rainbow.* In this variety the colours are red, blue, green, and yellow, on a grey-ground. It is more beautiful and brighter than the preceding. The radiated grey antimony-ore of Felsobanya in Hungary, and the specular iron-ore or iron-glance of Elba, are often beautifully iridescent.

                              γ. Columbine

γ. *Columbine* or *pigeon-neck tarnish.* The colours are the same as in the preceding, with this difference, that the tints of colour are paler, and the red predominates. Examples, native bismuth of Schneeberg.

δ. *Tempered-steel tarnish.* It consists of very pale blue, red, green, and very little yellow, on a grey ground. Example, grey cobalt-ore.

VIII. THE PERMANENT ALTERATIONS.

These must not be confounded with the tarnished colours. The tarnish occurs only on the surface ; the permanent alteration, on the contrary, proceeds by degrees through the whole mass of the mineral. This change takes place more or less rapidly in different minerals. The colours either become paler, when they are said to *fade,* or they become darker, and pass into other varieties. Thus chrysoprase, rose quartz, and red cobalt-ochre become paler ; whereas sky-blue fluor-spar becomes green, pearl-grey corneous silver-ore sometimes changes to brown, and lastly into black, and blue iron-earth changes from white, through different varieties of blue, to indigo-blue.

L   Utility

UTILITY OF COLOUR, AS A CHARACTER FOR DISCRIMINATING
NATURAL BODIES.

The older, and some of the modern mineralogists, as
we have already remarked, in their descriptions of the species of minerals, use only single varieties of colour, without attending to their natural relations; hence, as employed by them, colour is a character of but little value.
It was WERNER who first made the remark, that single
varieties are not characteristic, and that it is only by using
the whole range or suite of colour of the mineral, that we
are enabled to employ this character with advantage, in
the discrimination of the species that occur in the mineral
kingdom. Thus, it is not sufficient to say that epidote
is green, that beryl is green, or that topaz is yellow; we
must mention every variety of colour which these minerals possess, because each species of mineral is characterised by a particular suite or group of colours. Even
if different species should exhibit precisely the same
group of colours, still the character does not thereby
lose of its value and importance, because in the different
species, the colour suites would be associated with different
groups of external characters.

Although colours are frequently employed by botanists
for distinguishing species of plants, particularly in the
class Cryptogamia, still they in general hesitate in employing

ploying them in the discrimination of plants in the higher
divisions of the system.

It is alleged that the colours of plants change very
readily, particularly when cultivated in our gardens, and
that, therefore, so variable a character, should not be attended to. It is not denied, that the colours of plants
frequently undergo very considerable changes, when cultivated in our gardens; but these domesticated plants
are no longer the natural unaltered species, and therefore
are not objects of the attention of the systematic botanist. It is also known, that plants even in their natural
situations, owing to disease, experience great changes in
their colours; but these diseased individuals would sure-
ly never be taken by the botanist for characteristic examples of the species. Indeed it is highly probable, that
every species of plant in its natural region, has a determinate colour, or suite of colours. Hence colour may
be used as a most interesting character, particularly in
those systems of botany which are termed *natural*.

This character may also be advantageously used in giving
correct ideas of the changes of colour which plants experience by cultivation, or when removed from their natural
soil and climate. These changes have probably determinate
ranges in each species; thus, some run through certain
red and blue varieties, others through red and yellow, and
some through white, red and grey; and in others, the
change does not extend beyond varieties of one colour.
Some colours of the suite or range will be more prominent, more fixed, or more frequent than others; and the
extent of the colour suites will depend on the degree
of change in the situation, soil, and climate of the plant.
Interesting coloured maps might be constructed, to shew
the general changes in the colour of the vegetable world,
from

from the Equator towards the Poles; and the difference
of colours in vegetables in the two Hemispheres, and in
the Old and New World, might be delineated in a similar manner.

In the animal kingdom, the number of different colours is very great. They often form the most striking
feature in the external appearance of the species, and
hence have been considered by systematics as affording
discriminating characters of much value. The agricul-
turist, engaged in the breeding of animals, often witnes-
ses striking changes in their colours, and these varieties
of colour, either alone; or conjoined with other characters,
characterise his different breeds. But here, as in botany,
a regular systematic nomenclature of colour is much
wanted. To render the character of certain value in
botany and zoology, there ought to be established a number of fixed or standard colours, to which all the others
could be referred: The varieties should be defined
and arranged according to their resemblance to these
standard colours; and, lastly, the whole ought to be disposed in a regular and systematic order. The various
changeabilities of colour, their patterns or delineations,
and other similar varieties of character, ought to be defined and arranged as they are in mineralogy.

The anatomist will find it to his advantage, to use in
his descriptions some regular and fixed standard of colours; and in morbid anatomy, in particular, the importance of such an aid will be immediately perceived: Thus,
the various changes in the animal system, from the slight-
est degree of inflammation to complete gangrene, are
strikingly marked by the different colours the parts
assume. Accurate enumerations of these colours as they
occur

occur in single varieties, or in groups, conjoined with descriptions of the changes in form, transparency, lustre, consistency, hardness, structure and weight, observable in the diseased parts, will convey an accurate conception of the diseased part to those who have not an opportunity of seeing it. But to effect this, the anatomist and surgeon must agree on some fixed nomenclature, not only of colour, but also of form, transparency, lustre, consistency, hardness and structure ; and a better model cannot be pointed out, than that contrived by WERNER for the description and discrimination of minerals.

Lastly, the chemist will have daily opportunities of experiencing its utility ; and the meteorologist, and the hydrographer, by the use of an accurate and standard table of colours, will be enabled, in a much more satisfactory manner than heretofore, to describe the skies, and meteors of different countries, and the numerous varieties of colour that occur in the waters of the ocean, of lakes and rivers.

Many attempts have been made to delineate the different colours that occur in the mineral kingdom, with the view of enabling those who do not possess a mineralogical collection, or who may not be familiar with colours, to know the different varieties mentioned in the description of mineralogists. WIEDEMANN, ESTNER, LUDWIG, and several others, have published tables of this kind ; but all of them were deficient, not only in accuracy, but also in durability. Having the good fortune to possess a Colour-Suite of Minerals, made under the eye of WERNER, by my late friend H. MEUDER of Freyberg, and being desirous of making this collection as generally useful as possible, I mentioned my wish to Mr. SYME,

Painter

Painter to the Wernerian and Horticultural Societies, who readily undertook to make a delineation of all the varieties in the collection. This he executed with his usual skill and accuracy ; adding, at the same time, to the series several other colours, which he has distinguished by appropriate names, and arranged along with those in the Wernerian system. The whole have been published in a series of Tables, in a Treatise *, which ought to be in the hands of every mineralogist, and indeed in the possession of naturalists of every description.

II. THE COHESION OF THE PARTICLES.

What is meant by the cohesion of the particles of a mineral is understood by every one ; we shall here, therefore, only define their different states of aggregation.

Minerals are divided into *Fluid* and *Solid*, and the solid into

A. *Solid* in a stricter sense, and

B. *Friable*.

By Solid in a stricter sense, we understand that state of aggregation in which the particles cannot be overcome by the pressure of the finger ; and by Friable, that state of coherence which can be overcome by the simple pressure of the finger.

PARTI-

---

* " Werner's Nomenclature of Colours, with Additions, arranged so as to render it useful to the Arts and Sciences ; with Examples, selected from well known objects in the Animal, Vegetable and Mineral Kingdoms," by P. Syme, &c.Edinburgh, 1814.

PARTICULAR GENERIC EXTERNAL CHARACTERS.

I.

PARTICULAR GENERIC EXTERNAL CHARACTERS OF *SOLID* MINERALS.

CHARACTERS FOR THE SIGHT.

1. *THE EXTERNAL ASPECT.*

THE External Aspect of a mineral is that outline or contour which it has received from nature. Thus, if we have a piece of lead-glance, as it has been found loose, or imbedded in another mineral, we name the surface which it has received from nature, its Aspect. All those characters which we can discover by the eye, on this outline, are denominated the External Aspect of the mineral. They are of three kinds : 1 The External Shape : 2. The External Surface ; and, 3. The External Lustre.

1. THE EXTERNAL SHAPE
Is divided into four classes,

1. Common
2. Particular
3. Regular   } External Shape.
4. Extraneous

All of these classes have their subordinate differences, which we shall now describe ; and,

1. Common

1. *Common External Shape.*

Common External Shapes are those in which there are neither a determinate number of planes meeting under determinate angles, nor any resemblance to known natural or artificial bodies. As they occur more frequently than the other shapes, they are named Common External Shapes.

Six different kinds are enumerated by Werner, which are distinguished according to their relative length, breadth, and thickness, their relative magnitude, and their connections with other minerals. The kinds are, *massive, disseminated, in angular pieces, in grains, in plates,* and *in membranes.*

A. *Massive,* is that common external shape which is from the size of a hazel-nut to the greatest magnitude, and whose dimensions in length, breadth, and thickness, are nearly alike. It occurs imbedded in other minerals, and it is intermixed with them at their line of junction. Examples, galena or lead-glance and copper-pyrites.

B. *Disseminated,* is from the size of a hazel-nut until it is scarcely visible, and its dimensions in length, breadth and thickness are nearly alike. It is imbedded, and is intermixed with the inclosing mineral at the line of junction. It is divided into

a. *Coarsely disseminated,* which is from the size of a hazel-nut to that of a pea. Examples, copper-pyrites and brown-spar.

b. *Minutely disseminated,* from the size of a pea to that of a millet seed. Example, tin-stone in granular quartz.

c. Finely

c. *Finely disseminated*, from the size of a millet
seed until it is scarcely visible. Example,
brittle silver-glance in brown-spar.

C. *In angular pieces.* Minerals having an angular
shape, in which the length, breadth and thickness are nearly alike, which are found loose, or
slightly imbedded, and without any intermixture
with the inclosing mineral at the line of junction, and from the size of a hazel-nut and upwards, are said to occur in angular pieces. It is
distinguished from the massive by its occurring
either loose, or not intermixed with the basis at
the line of junction. Of this external shape there
are two kinds.

a. *Sharp cornered*, as in quartz and calcedony.

b. *Blunt-cornered*, as in common opal.

The sharp-cornered occur but rarely; opal sometimes exhibits this form. The blunt-cornered
are much more frequent, and are either original, or have been formed by attrition, when
they are named *pebbles*, or *rolled pieces*. The
original pieces are distinguished by a surface
which is pretty smooth and shining, and marked with numerous angles and hollows; while
the pebbles or rolled pieces have an even and
rough surface, and approach more to the
roundish form. Calcedony affords a good example of the first kind, and the rolled pieces
of common quartz, rock-crystal and flint, found
in the beds of rivers, of the second kind.

D. *In grains.* Minerals which are either loose, or
slightly imbedded, not being intermixed with the
basis in which they are imbedded, and not larger

M   than

than a hazel-nut, are said to occur in grains.
This shape is distinguished

a. With regard to size, into

α. *Large*, that is, when they are from the size of
a hazel-nut to that of a pea. Examples,
meadow-ore and precious garnet.

β. *Coarse*, from the size of a pea to that of a
hemp-seed. Example, pyrope.

γ. *Small*, from the size of a hemp-seed to that of
a millet-seed. Examples, precious garnet,
pyrope and iron-sand.

δ. *Fine*, from the size of a millet-seed until it becomes nearly undistinguishable. Example,
platina.

The grains are further distinguishable

b. With regard to the exacter determination of the
shape into

α. *Angular grains*, as in iron-sand.

β. *Flattish grains*, as in platina and gold.

γ. *Roundish-grains*, as in pyrope and precious
garnet.

c. With regard to connection with other minerals.

α. In loose grains.

β. In imbedded grains.

γ. In superimposed grains.

E. *In plates.* Minerals which occur in external shapes
whose length and breadth are great in comparison of their thickness, in which the thickness is
not equal throughout, and is so considerable, as
to allow the fracture to be distinguished, are said
to occur in plates. The maximum thickness of
plates

plates is half an inch. According to the degrees
of thickness, it is distinguished into

a. *Thick plates*, when they are as thick as the back
of a penknife. Example, red silver-ore.

b. *Thin plates.* Where thinner than the preceding.
Example, silver-glance, or sulphureted silver-
ore.

F. *In membranes or flakes.* This shape is distinguished from the former by its thinness, as it never
greatly exceeds the thickness of common paper,
and the fracture cannot be discriminated. Its
different degrees of thickness are,

a. *Thick.* Example, silver-glance or sulphureted
silver-ore.

b. *Thin.* Example, iron-pyrites.

c. *Very thin.* Example, copper-pyrites on clay-
slate.

2. *Particular External Shape.*

Particular external shapes differ from the common external shapes, in bearing a resemblance to natural or artificial bodies, and in being far more characteristic and
varied in their aspect. They are called *particular*, because
they are not so common among minerals, as the common
external shapes. They are divided into five classes, entitled, *longish, roundish, flat, cavernous,* and *entangled.*
Each of these species have their subordinate kinds, which
we shall now describe.

A. *Longish Particular External Shapes.*

a. *Dentiform,* adheres by its thick extremity, and
becomes gradually thinner, incurvated, and at
length terminates in a free point, so that it resembles a canine tooth, whence its name. Its
length

length is from a quarter of an inch to a foot.
It is one of the rarer kinds of external shapes,
and is peculiar to certain metals. Examples,
native silver and silver-glance, or sulphureted
silver-ore, and native copper.

b. *Filiform,* adheres by its thicker extremity, and
terminates by an almost imperceptible diminution of thickness, and is usually curved in different directions. It is thinner and longer than the
dentiform. Examples, native silver, silver-glance
or sulphureted silver-ore, native gold, and native
copper.

c. *Capillary.* When the filiform becomes longer and
thinner, it forms the capillary. It is generally
much entangled, and sometimes the threads are
so near each other that it passes into the compact. Example, native silver.

d. *Reticulated* is composed of many straight threads,
which are sometimes parallel and sometimes meet
each other at right angles, and form a net-like
shape. The whole is a series of minute crystals,
and is distinguished from the capillary by its
threads being always straight. Examples, native silver, native copper, and copper nickel.

e. *Dendritic.* In this external shape we can observe
a trunk, branches, and twigs, which are distinguished from each other by their thickness, the
trunk being the thickest. It is divided into *regular* and *irregular dendritic;* in the first, the
branches are set on the trunk, and the twigs on
the branches at right angles, or at angles more
or less acute, as in native silver; in the second,
the branches proceed from the stem, and the
twigs

twigs from the branches irregularly, and the shape is not, as is the case with the regular den- dritic, a series of crystals. Examples, native copper and brown hematite.

f. Coralloidal or *coralliform.* When two or three branches, having rounded or pointed extremities, proceed from one stem, the coralloidal external shape is formed. There are usually many stems together. From its resemblance to coral, it is denominated coralloidal. The variety of arrago- nite, called *flos ferri,* is an excellent example of this kind of particular external shape.

g. Stalactitic. A mineral is said to possess a stalac- titic external shape when it consists of different straight more or less lengthened rods, which are thickest at their attachment, and become nar- rower at their free extremity, which is rounded or pointed. Examples, calc-sinter, brown he- matite, and calcedony.

h. Cylindrical consists of long, rounded, straight, im- perforated, usually parallel rods, which are at- tached at both extremities, and are generally thicker at the extremities than the middle. The interstices are either empty, or filled up with an- other mineral. Examples, galena or lead-glance and brown ironstone.

i. Tubiform consists of long, usually single, perfora- ted tubes, which are somewhat longitudinally knotty. Example, calc-sinter.

k. Claviform is the reverse of stalactitic; it is compo- sed of club-shaped parallel rods, which adhere by their thin extremities. Example, compact black ironstone.

l. Fruc-

l. Fructicose. This external shape is formed when many branches issue from a common stem and meet together partywise, so that the whole when viewed from above has a fructicose aspect, not unlike the appearance of colewort. Examples, calc-sinter and black ironstone.

B. *Roundish Particular External Shapes.*

    a. Globular. Under this, as mentioned in the tabular view, are comprehended,

      α. Perfect globular or *spherical,* as in alum-slate and pisiform iron-ore.

      β. Imperfect globular, as in calcedony, carnelian, agate, and iron-pyrites.

      γ. Ovoidal or *elliptical.* Example, rounded masses of quartz in puddingstone.

      δ. Spheroidal. When the spherical is compressed the spheroidal is formed. Examples, Egyptian jasper and calcedony.

      ε. Amygdaloidal. When the ovoidal is compressed in the direction of its length, the amygdaloidal is formed. Examples, zeolite, calcareous-spar and green earth.

    b. Botryoidal consists of large segments of small balls, which are regularly heaped together, and have many interstices. It resembles grapes, whence its name. Examples, black ironstone and calcedony.

    c. Reniform consists of small segments of large balls, which are so closely set together that no interstices are formed. Examples, red hematite, calcedony, and malachite.

                          d. Tuberose.

d. Tuberose. This shape consists of irregular roundish or longish elevations and depressions. Examples, flint and menilite.

e. Fused-like, or *liquiform* *. It consists of numerous
very flat roundish elevations, which are generally depressed in the middle. The whole has a
rough and glimmering surface, and resembles the
surface of slowly cooled metal. It is one of the
rarest of the external shapes, and has been hitherto found but in one mineral, that is galena
or lead-glance ; which variety is found in the
mine called Alten Grünen Zweig, situated behind the mining village of Erbisdorf near Freyberg, and in that named Methusalem in the same
neighbourhood.

C. *Flat Particular External Shapes.*

a. Specular has on one side, seldom on two opposite
sides, a straight smooth shining surface. It occurs in veins. Examples, galena or lead-glance,
copper-pyrites, red ironstone, and quartz.

b. In leaves. In this external shape there are thin
leaves, which are either irregularly curved, or
are straight, and have throughout the same
thickness, It is distinguished from the external
shape in membranes by the uniformity of its
thickness, by its irregular curvatures, its continuity, (the membranous external shape being
often interrupted,) and its usual adherence by
one

---

* The term *liquiform* is used by Mr WEAVER, in his excellent translation of WERNER's Treatise on the External Characters of Minerals.

one extremity, shewing that it is a kind of crys
talline shoot. It occurs frequently in native gold,
but seldom in native silver.

D. *Cavernous Particular External Shapes.*

    a. *Cellular.* A mineral is said to be cellular, when it
is composed of straight or bent tables, which
cross together in such a manner as to form empty spaces or cells.

    α. *Straight*, or *angulo-cellular*, in which the tables
are straight. It is divided into

    1. *Hexagonal*, as in quartz, and cellular pyrites.
    2. *Polygonal*, as in quartz.

    β. *Circulo-cellular*, in which the tables are curved.
It is divided into

    1. *Parallel*, where the cells are in rows, and of a
cylindrical shape. Example, quartz.

    2. *Spongiform.* In this figure the cells are cy-
lindrical, of equal size, very small, bent, ly-
ing near each other, disposed in different
directions, and not parallel as in the preceding. Example, quartz.

    3. *Indeterminate.* In this figure the cells have no
particular shape, and are of different sizes.
Example, compact ironstone.

    4. *Double circulo-cellular*, consists of large flat
cells, whose walls are beset with other
smaller ones. Example, quartz.

    b. *Impressed.* That is, when a mineral shews the
impression of any particular or regular external
shape of another mineral. It borders on the cellular shape, and is formed when a newer mineral is deposited over an older, the form of which

    it

it assumes, and retains even after the impressing
mineral has been destroyed or removed.

a a. With impressions of crystals.

   α. Cubical, as in quartz or iron-pyrites from fluor-
spar.

   β. Pyramidal, as in hornstone, originating from
calcareous spar.

   γ. Tabular, as in quartz, originating from heavy
spar.

b b. With impressions of particular external shapes.

   α. Conical, in native arsenic.

   β. Globular, in silver-glance or sulphureted silver-ore, from red silver-ore.

   γ. Reniform, in silver-glance or sulphureted silver-ore, from red silver-ore.

c. Perforated, consists of long vermicular cavities which
occupy but an inconsiderable portion of the mass,
and terminate on the surface in small holes. When
the holes become very numerous, it passes into
spongiform. Example, bog iron-ore.

d. Corroded. A fossil is said to be corroded when it is
traversed with numerous hardly perceptible roundish holes. The volume occupied by the holes is
nearly equal to that of the basis. It has the appearance of wood which has been gnawed by insects. Examples, quartz, galena or lead-glance,
and silver-glance or sulphureted silver-ore.

e. Amorphous is composed of numerous roundish and
angular parts that form inequalities, between which
there are equally irregular hollows. The whole has
the appearance as if a number of small balls and

              N   angular

angular pieces were heaped on one another. Examples, silver-glance or sulphureted silver-ore, and
meadow-ore.

f. Vesicular. When a mineral has distributed through
its interior many single, usually round, elliptical,
and spheroidal, also amygdaloidal, or irregular
shaped cavities, it is said to be vesicular. The cavities are usually less in volume than the solid part
of the mineral, and they are larger than the holes
or cavities in the corroded external shape. Examples, wacke and lava.

E. *Entangled Particular External Shape.*

   a. Ramose, It is composed of longish, angular, more
   or less thick branches which are bent in different
   directions, but in which no trunk or common stem
   is to be observed. It probably originates from
   the greater magnitude of the vesicles in the vesicular, the vesicles breaking into each other. Examples, meteoric-iron, silver-glance or sulphureted silver-ore, and native copper.

3. *Regular External Shape, or Crystallization.*

Every external shape, whose natural contour or outline
is composed of a determinate number of planes, which
meet together in a determinate manner, is denominated
a Crystal.

In describing crystals, we have to consider, *A. Their
Genuineness; B. Their Shape; C. Their Attachment;
D. Their Magnitude.*

A. The

A. The Genuineness of Crystals.

This refers to the division of crystals into *True* and *Supposititious*. The *true* are the forms which the same substance always assumes, and which are peculiar to it; the *supposititious* are those regular figures whose shape does not depend on the substance of which they are composed, but is owing to pre-existing crystals, or crystal-moulds.

Supposititious Crystals are formed in two ways :

1. When an imbedded crystal falls out and leaves an empty mould, which is afterwards filled up with fossil matter, a figure or crystal corresponding in shape to the mould is formed. The supposititious crystals formed in this manner are smoother, and have sharper edges and angles than the succeeding kind, and their interior is often hollow and drusy.

2. When a mineral is deposited over a pre-existing crystal, and assumes its figure, the second kind of supposititious crystal is formed. The pre-existing crystal either remains, forming the nucleus, or it disappears when the supposititious crystal is hollow. It differs from the first kind in having generally a rough and drusy surface, blunter edges and angles, and the inner surfaces smooth.

The first kind of supposititious crystal is a cast or filling of the space formerly occupied by true crystals ; the second is merely an incrustation of true crystals.

True

True and supposititious crystals are distinguished from each other by the following characters :

a. *True crystals.*

　α. Are transparent and semitransparent.

　β. Their planes are smooth and shining or splendent, or they are regularly streaked.

　γ. Their angles and edges are sharp.

　δ. They form particular characteristic suites.

b. *Supposititious crystals.*

　α. The planes are never smooth and shining, or regularly streaked; on the contrary, are generally rough and dull.

　β. The angles and edges are not so sharp as in true crystals, but are generally somewhat rounded.

　γ. They are usually hollow, and their internal surface is drusy.

　δ. They are not like true crystals, connected by transitions with other crystals of the same species; Thus the octahedral supposititious crystals of quartz, which originate from fluor-spar, do not belong to the suite of quartz.

　ι. Even in their internal structure they are different from true crystals ; for they seldom present a fracture inclining to foliated.

The following are well known instances of supposititious crystals.

　　1. Octahedral crystals of quartz, originating from fluor-spar.

　　2. Cubic crystals of quartz, from fluor-spar.

　　3. Flint in double three-sided pyramids, from calcareous-spar.

4. Quartz

4. Quartz in oblique four-sided tables, from heavy-spar.

B. The Shape of Crystals.

The shape of crystals is determined by the number and form of the planes or faces, and the edges and angles which form the contour or outline * Amidst the great variety of crystals that occur in the mineral kingdom, there are some simple ones, which are composed of but few planes, that do not vary much in shape ; and of others, in which the planes are not only numerous, but present great differences in form. These simple forms are nearly allied to the more complex ones, and gradually pass into them by a change in the shape of their planes. On this circumstance WERNER has founded a Crystallographic System, remarkable for its simplicity, and the ease with which it enables us to acquire distinct conceptions of the most complicated crystallisations. He considers these simple forms as the basis of the others, and names them *Fundamental Figures.* We can distinguish in them one, or, at the utmost, two sets of planes, which run in two directions, and inclose the crystal on all sides. The cube is an example of a' fundamental figure with one set of planes; the prism, pyramid and table, are examples of fundamental figures with two sets of planes, which are named *lateral* and *terminal planes.* All those crystals in which we observe many different sorts of planes, he considers as changed or altered fundamental figures ; and names the other planes which are generally smaller, and differ from the planes of

the

---

* When the faces are very small, they are named *facets.*

the fundamental figure in direction, and in being further removed from the centre of the crystal, *Altering Planes*. We have thus, according to this method, to consider, first, *The Fundamental Figures*, and then their *Alterations* or *Modifications*.

I. The Fundamental Figures.

The fundamental figures, as already mentioned, are composed of one or two sorts of planes. In order to discover these planes in the altered fundamental figures, we have only to conceive the planes that lie nearest the centre of the crystal, and which are generally the largest, extended on all sides until they join.

In the fundamental figure are observed and attended to, I. *Its Parts*. II. *Its Varieties or Kinds*. III. *The Differences of each Fundamental Figure in Particular*.

I. *Parts of the Fundamental Figure*.

The fundamental figure is composed of *lateral* and *terminal planes*; of *lateral* and *terminal edges*; and of *angles*.

1. *Lateral planes* are the greatest planes that bound the smallest exent. *Terminal planes* are the smallest planes that bound the greatest extent. In the prism they form the *bases*, but in the table they are the smaller planes that surround the two largest planes.

2. *Lateral edges* are formed by the junction of two lateral planes, as in the prism and pyramid; but in the table, where the lateral planes do not meet, the lateral edges are those formed by the meeting

ing of the lateral planes and the terminal planes ;
or we say, they are the edges of the lateral faces
of the table, so that there are eight lateral edges
in a four-sided prism, &c. *Terminal edges* are
formed by the junction of lateral and terminal
planes, as in the prism and pyramid ; or they are
those that surround the terminal planes in the
prism or the base of the pyramid: they are
also formed by the junction of two terminal
planes, as in the table *.

3. *Angles.* The point in which three or more planes
meet, is called a *solid angle.*

II. *The Varieties or Kinds of the Fundamental Figure.*

WERNER admits seven fundamental figures, *viz. icosahedron, dodecahedron, hexahedron, prism, pyramid, table,*
and *lens.*

1. *Icosahedron* is a solid having twenty equilateral
triangular planes, that meet together under nearly equal obtuse angles ; and of twelve angles, so
that there are always five planes to form an angle.
Fig. 1. Pl. I. It is rare. Example, iron pyrites.

2. *Dodecahedron* has twelve regular pentagonal planes
that meet under equal obtuse angles ; and of
twenty angles. Fig. 2. Pl. I. It occurs but seldom. Example, iron-pyrites.

3. *Hexahedron*

---

* The terminal edges in the table are these that measure its thickness.

3. *Hexahedron* is a solid, having of six quadrilateral
planes and eight angles. It includes the cube
Fig. 3. Pl. I. ; and the rhomboid, Fig. 4. Pl. I.
which is sometimes considered as a double three-
sided pyramid, in which the lateral planes of the
one are set on the lateral edges of the other.
It is very frequent. Examples, calcareous-spar
and fluor-spar.

4. *Prism* has an indeterminate number of quadrangular lateral planes terminated by two equal terminal planes parallel to each other, and having as
many sides as the prism has lateral planes. Fig. 5.
Pl. I. This is the most frequent of the fundamental figures. Examples, calcareous-spar, rock-
crystal, schorl and topaz.

5. *Pyramid* has an indeterminate number of triangular
lateral planes converging to a point, and of a base
possessing as many sides as the figure has lateral
planes. Fig. 12. Pl. I. The terminal point is
called the *summit-apex*, and the flat part the *base*.
It occurs very often. Examples, calcareous-spar
and amethyst.

6. *Table* has two equal and parallel lateral planes,
which are very large in comparison of the others,
and which are bounded by an indeterminate number of small four-sided terminal planes. Fig. 15.
Pl. I. It is but a very short prism. It is proper to observe, that the parts of the table are not
denominated as those in the prism, but inverse-
ly, the lateral planes of the table corresponding
to the terminal planes of the prism, and the terminal planes of the table to the lateral planes of
the

the prism. It does not occur very often. Examples, heavy-spar and calcareous spar.

7. *Lens* has two curved faces or planes. Fig. 19. and 20. Pl. I. It occurs but seldom. Example, sparry ironstone.

III. *The Differences of each Fundamental Figure in particular.*

Here we have to determine, 1. *The simplicity.* 2. *Number of Planes.* 3. *Proportional size of the planes to one another.* 4. *Direction of the Planes.* 5. *Angles under which the planes meet.* 6. *Plenitude, or fulness of the crystals.*

1. *Simplicity.*

With respect to simplicity, the fundamental figures are either *simple* or *double.* This distinction, however, is confined to the pyramid, as the other six kinds of primitive figures occur simple only. Fig. 12. Pl. I. is a simple pyramid ; and Fig. 13. Pl. I. a double pyramid.

The simple figure is also distinguished, in regard of its position, into *erect* or *inverted*, according as it adheres by its base or its summit. The inverted has hitherto occurred only in calcareous-spar, and is very rare.

In the double figure we have to attend to the placing of the lateral planes ; thus the lateral planes of the one pyramid are placed either *straight* or *oblique* on the lateral planes of the other pyramid. In Fig. 13. Pl. I. they are placed straight ; and in Fig. 14. Pl. I. they are placed obliquely ; or the lateral planes of the one pyramid are

O   set

set either on the lateral edges, as in Fig. 35. Pl. II. or on
the lateral planes of the other, as in Fig. 13. Pl. I.

2. Number of Planes.

The number of planes in the icosahedron, dodecahedron, hexahedron, and lens, is always determinate, but
in the prism, pyramid, and table, is indeterminate. In
the prism and pyramid, it is only the lateral planes that
vary in number, but in the table it is the terminal
planes.

The *prism* occurs with two, three, four, six, eight,
nine, and twelve lateral planes. The trihedral or three-
sided occurs in schorl and tourmaline. The four-sided
or tetrahedral prism, Fig. 5. Pl. I. occurs very often; we
have examples of it in felspar, zeolite, zircon, and
heavy-spar. The six-sided or hexahedral prism, Fig. 8.
Pl. I. occurs very often, and is the most common
prismatic crystallization; quartz, emerald, beryl, calcareous-spar, heavy-spar, and actynolite afford examples
of it. The octahedral, or eight-sided prism, is rare; it occurs in augite and topaz. The nine and twelve sided
prisms are merely varieties of the preceding figures; the
first is formed by the bevelling of the lateral edges of the
trihedral prism, the other by the truncation of the lateral edges of the six-sided prism. Beryl affords an
example of the twelve-sided, and tourmaline of the nine-
sided prism.

The pyramid occurs with three, four, six, and eight
sides. The three-sided pyramid, Fig. 9. Pl. I. is either
single or double; of the single we have examples in grey
copper-ore, spinel, copper-pyrites, and many other minerals. Examples of the second occur in calcareous-spar,

as

as in Fig. 10. Pl. I.   The four-sided pyramid is the
most common, and is always double, Fig. 11. j l. I. ;
when it appears single, the one half is either hid in part
or altogether in the matrix ; diamond, zircon, and fluor-
spar, are examples. The six sided or hexahedral pyramid occurs single, as in Fig. 12. Pl. I. and double,
as in Fig. 13. Pl. I.   Examples of it occur in sapphire
and calcareous-spar, red silver ore, white lead-ore, quartz,
and amethyst. The eight sided is always double and
acuminated on both extremities by four planes, as in
Fig. 35. Pl. II.   Examples of it occur in leucite, garnet, and silver-glance or sulphureted silver-ore.

   The table has four, six, or eight terminal planes. The
three-sided tables are mere varieties of some of the other
figures.   The four-sided table, Fig. 15. Pl. I. occurs
frequently, as in heavy spar, white ore of antimony, and
yellow lead ore. The six-sided table, Fig. 17. Pl. I. oc.
curs still more frequently ; we have examples of it in
mica, calcareous spar, heavy-spar, and native gold. The
eight sided table occurs in heavy spar and yellow lead-
ore.

3. *Proportional Size of the Planes to one another.*

   This character is not of very much importance. The
planes are either equilateral or unequal ; where they are
unequal, they are either indeterminately or determinately unequal. The determinately unequal planes are, as
mentioned in the tabular view, *a*. Alternately broad and
narrow ; *β*. With two opposite planes broader ; *γ*. With
two opposite planes narrower. We shall illustrate this
character by examples drawn from the fundamental
figures.

In

In the hexahedron, dodecahedron, and icosahedron, the planes are alike; when any dissimilarity occurs, it is merely accidental, and is therefore indeterminately unequal. The three-sided prism shews only slight indeterminate inequalities. The four-sided prism is not always equilateral; sometimes two opposite planes are broader than the others, when the prism is said to be broad, as in zeolite. The six-sided prism is almost always equilateral; its varieties are generally accidental, excepting the following, which are somewhat characteristic. 1. The two opposite lateral planes broader than the others, as in actynolite and heavy-spar. 2. The planes alternately broader and narrower, as in calcareous-spar. The eight and nine sided prisms afford only accidental or indeterminate varieties, as augite, topaz, and tourmaline.

In the pyramid, sometimes the two opposite planes are larger than the others, when it is said to be broad.

The four-sided table is usually equilateral; it has sometimes, however, two opposite lateral planes longer than the others, as in Fig. 16. Pl. I. The six-sided table is sometimes unequilateral, or two opposite planes are larger than the others, as in Fig. 18. Pl. I.; and the eight-sided table is usually longish.

4. *The Direction of the Planes or Faces.*

The direction of the planes or faces is either *Rectilinear* or *Curvilinear*.

Rectilinear is the most common, and is the case with almost all the fundamental figures.

Curvilinear,

Curvilinear planes * differ partly by the *position* of the
curvature, which is either *concave*, as in fluor-spar ; *convex*, as in diamond ; *concavo-convex*, as in sparry ironstone;
saddle-shaped, as in the lens ; they differ also by the
shape, being either *spherical*, as in brown-spar ; *cylindrical*, in which the convexity is either parallel with the
sides, as in iron-pyrites, or parallel with the diagonal, as
in fluor-spar ; and *conical*, as in gypsum, and probably
also in galena or lead-glance.

5. *Angles under which the Planes meet.*

The size of the angles formed by the meeting of the
planes, is determined either by means of an instrument
named *Goniometer* or angle-measurer, or simply by ocular
inspection. Several different kinds of goniometer have
been contrived. The first invented, and that which
is at present most generally used, is represented in
Plate VII. It consists of a semicircle of brass, divided
into degrees. At its centre C is fixed a pin, upon which
slide the two arms AB and GF. The last of these GF,
by means of a screw, may be fixed in any position, so
that the distance between the end G and the centre, may
correspond with the face of the crystal to be measured.
The other arm AB is drawn up, till the distance between
B and the centre corresponds as nearly as possible with
the size of the other face of the crystal. It is then turned round, till the angle of the crystal to be measured,
corresponds

---

* It is not geometrically correct to speak of curved planes ; yet, as the
term plane is more generally used by mineralogists than face or side, I
have not thought it necessary to make any alteration.

corresponds exactly with the angle B c G; the arm AB
then cuts the same circle in the angle which corresponds
with that of the crystal. There is a hinge upon the middle of the brass semicircle which is not seen in the figure.
By means of it, one-half of the semicircle may be thrown
back, when the crystal to be examined happens to be so
situated, in a group, that the arms could not otherwise
come at it. This is the goniometer used by ROME DE
LISLE, HAUY, BOURNON, BERNARDI, WEISS, and other
crystallographers, and which is sufficiently accurate for the
general practice of mineralogy Doctors WOLLASTON and
BREWSTER have each published descriptions and figures
of goniometers, contrived on optical principles, and susceptible of greater accuracy, than the instrument already
described. Dr WOLLASTON's instrument is drawn and described in the *Philosophical Transactions* for 1809,
p. 11.; and that of Dr BREWSTER in his *Treatise on New
Philosophical Instruments*, p. 89.

The other mode of determining the magnitude of crystals, namely, by ocular inspection, without the aid of the
goniometer, is that practised by WERNER. In this way
he determined the whole of the species in the system, and
it is known that it has enabled him to do so with accuracy.

Several different kinds of angles occur in the fundamental figures : these are the *angles of the lateral edges,
angles of the terminal edges*, and *the summit angles*.

1. The angles formed by the meeting of the lateral
   planes, are named the *angles of the lateral edges,*
   or, to shorten the description, simply, *lateral
   edges*. Thus we say, acute and obtuse lateral
   edges, in place of acute and obtuse angles formed

ed by the meeting of the lateral planes. The lateral edges are either *equiangular*, or *unequiangular*. In the icosahedron, all the edges are *equiangular*. In the dodecahedron, the edges are
equiangular. The hexahedron is either *equiangular* and also *rectangular*, or *unequiangular* and
oblique angular. The rectangular hexahedron
is named *cube*; the oblique-angular, *rhomboid*. In the prism, the lateral edges are either
equiangular or *unequiangular*. The four-sided
prism, with unequiangular lateral edges, is denominated an oblique four-sided prism, as Fig. 6.
Pl I. In the pyramid, the lateral edges are generally equiangular ; seldom unequiangular. The
same is the case with the table ; when the edges
are unequiangular, we say the terminal planes
are set obliquely on the lateral planes.

2. The *terminal edges* are either *equiangular* or *unequiangular*. In the prism they are generally
equiangular, as in Fig. 6. Pl. I ; and sometimes
unequiangular, when we say that the terminal
planes are set obliquely on the lateral planes, as
in Fig. 7. Pl. I. They are always equiangular
in the pyramid. In the table they are as in the
lateral edges of the prism.

3. The *summit angle*. It occurs only in the pyramid.
It is measured from plane to plane, or from plane
to edge Werner determines it in degrees in
the following manner.

a. Extremely acute is from 1° to 30°.

b. Very acute from 30° to 50°. Example, sapphire.

c. Acute from 50° to 70°. Example, calcareous spar.

d. Rather

d. Rather acute from 70° to 90°. Example, quartz.

e. Rectangular 90°. Example, zircon.

f. Rather obtuse, or rather flat, from 90° to 110°. Example, honeystone.

g. Obtuse or flat from 110° to 130°. Example, calcareous-spar.

h. Very obtuse or very flat from 130° to 150°. Example, tourmaline.

i. Extremely obtuse, or extremely flat, from 150° to 180°

6. *Plenitude or Fulness of the rystals.*

A. Full, as in almost all crystals.

B. Excavated at the extremities, as in green lead-ore.

C. Hollow. Olive-green-coloured calcareous-spar, from Schemnitz in Hungary, occurs in acute hollow three-sided pyramids.

II. The Alterations on the Fundamental Figure.

These occur on the edges, angles and planes, and are produced by i. *Truncation;* ii. *Bevelment;* iii. *Acumination; and,* iv. *Division of the Planes.*

i. *Truncation.*

When we observe on a fundamental figure, in place of an edge or angle, a small plane, such a plane is denominate a Truncation

These

---

* Werner, in assuming certain fundamental figures, and supposing them variously modified, does not propose to point out the course followed

by

These new planes are named *Truncating Planes*, and
the edges which they form with the other planes *Truncating Edges.*,

We have here to observe what relates to the *situation,
magnitude,* the *setting on* or *position,* and the *direction* of
the *truncation.*

  a. In regard to the *situation* of the truncation, it is
either on the edges or on the angles, and sometimes a few, sometimes all the angles and edges
of the figure are truncated. Fig. 21. Pi II. is
a cube truncated on the angles; and Fig. 22. a
cube truncated on the edges.

  b. In regard to the *magnitude* of the truncation, it is
either *deep* or *slight,* according as more or less of
the fundamental figure is wanting; and consequently the truncating planes are proportionally
greater or smaller.

  c. The planes are *set on* either straight or oblique.
They are said to be *set on straight,* when they are
equally inclined on all the adjacent planes; and
set on obliquely, when they are not equally inclined on the adjacent planes.

  d. The truncating planes in regard to their *direction,*
are either *straight* or *curved.* In the latter case,
we also say that the edge or angle is *rounded off.*

       P   ii. *Bevelment*

---

by nature in their formation. He employs a peculiar descriptive language
to convey a conception of their forms, not to explain the order of their construction. When he describes a crystal as truncated on its angles or edges,
he knows very well that nature does not begin by making a crystal complete, in order afterwards to truncate it more or less on one or other of its
parts, he only expresses by this term the appearance the crystal presents to
the eye, thus employing a well known term to express an operation of nature which still remains to us a mystery.

ii. *Bevelment or Cuneature.*

When the edges, terminal planes, or angles, of a fundamental figure are so altered, that we observe in their place two smaller converging planes, terminating in an edge, it is said to be bevelled. These two newer or additional planes are named *bevelling planes ;* and the edge formed by their meeting, the *bevelling edge.* We have here, again, to observe the *situation, magnitude, angle, uniformity* and *setting on* of the bevelment.

   a. In regard to *situation,* the bevelment is generally on the edges, sometimes on the terminal planes, and seldomer on the angles. Fig. 23. Pl. II. is a cube bevelled on the edges ; Fig. 24. a three-sided prism, bevelled on the lateral edges ; Fig. 25. a four-sided prism, bevelled on the terminal planes, the bevelling planes set on the lateral edges ; Fig. 26. a table bevelled on the terminal planes ; and Fig. 27. an octahedron bevelled on all the angles.

   b. In regard to *magnitude* of the bevelment, it is either *deep* or *slight,* according as more or less of fundamental figure is wanting.

   c. In regard to the *angle,* the bevelment is *obtuse* or *flat,* or *rectangular,* or *acute angular.*

   d. The bevelment in regard to *uniformity,* is either *unbroken,* when it extends in one direction ; or *broken,* when each bevelling plane consists of several planes,—sometimes of two planes, when it is said to be *once broken,*—and sometimes of three planes, when it is said to be *twice broken.*

   e. In regard to the *setting on,* we have to attend to the *position* and the *direction* of the bevelment.

a. In

a. In regard to the *position*, the bevelment is either on planes or on edges.

b. The *direction* varies only when the bevelling planes are set on the terminal planes. It is said to be *set on straight* when it is at right angles to the axis of the crystal; and *set on oblique*, when it forms an oblique angle with the axis of the crystal.

iii. *Acumination*.

A fundamental figure is said to be acuminated when, in place of its angles or terminal planes, we find at least three planes which converge into a point, and sometimes, but more rarely, terminate in an edge.

We have here to observe the parts of the acumination ; these are,

The acuminating planes.

The edges of the acumination, which are,

The proper acuminating edges, those formed by the meeting of acuminating planes.

The edges which the acuminating planes make with the lateral planes of the fundamental figure.

The terminal edges of the acumination, which are formed by the terminating of the acuminating planes in an edge or line.

The angles of the acumination ; which are,

The angles which the acuminating planes form with the lateral planes ; and,

The summit angle.

We have to determine in the acumination, its *situation;* the *number of its planes ; proportional magnitude of the planes among themselves ;* the *setting on* or *application of*
the

the planes ; the *angles of the acumination* ; its *magnitude* and *termination.*

a. *Its situation* is either on angles, as in Fig. 28. and 29. Pl. II. or on terminal planes, as in Fig. 30. Pl. II.

b. *The number of planes* is three, as in Fig. 33. and 34. Pl. II.; four, as in Fig. 30. Pl. II.; six, as in Fig. 31. Pl. II.; or eight, as in Fig. 35. Pl. II.

c. *The proportional magnitude of the planes among themselves,* is a character of but little importance., They are generally determinately unequal, as in heavy-spar ; or undeterminately unequal, as in rock-crystal.

d. *The setting on* or *application of the planes* refers to their position on edges, as in Fig. 29. 32. 34. Pl. II. or planes, as in Fig. 28. 30. 31. and 33. Pl. II. When the planes of an acumination are not set on all the edges or planes of the fundamental figure; but only on the alternate planes or edges, it is said to be *set on alternately,* as in Fig. 33. and 34. Pl. II. ; and when the acuminating planes on both extremities of the fundamental figure are set on the same planes or edges, it is said to be *conformable* (rechtsinnig), as in Fig. 31. Pl. II. ; but when the planes on opposite ends of the figure are set on different planes or edges, it is said to be *unconformable,* as in Fig 33. and 34. Pl. II. The same expressions are applied to alternate Truncations.

e. The *angle of the acumination,* or the *summit angle,* is either obtuse or flat, as in garnet ; rectangular, as in zircon ; or acute, as in calcareous-spar.

f. The *magnitude ;* according to which crystals are *deeply acuminated,* as in cubic crystals of fluor-spar, whose angles are acuminated with six planes ;

planes; *slightly acuminated,* as in copper-pyrites
or grey copper-ore.

g. The *termination,* according as the acumination
ends in a point, which is the usual mode, or in a
line or edge, which is less frequent.

In order to form a more distinct idea of truncation,
bevelling and acumination, let us take a cube, prism, pyramid, or any other perfect fundamental figure represented in wood, and cut off each of the edges or angles at one
stroke, so that in its stead a plane shall appear; this will
be Truncation. But if the extreme planes, the edges, or
the angles of any of these fundamental figures be cut off
with two converging strokes, the one from this side, the
other from that, so that two planes arise, which, terminating in a line, shall present an edge; this will be Bevelling. And if the extreme planes or the angles be cut
off at several strokes, all converging together, so that
more than two plains arise, commonly terminating in a
point, we shall obtain Acumination.

iv. *The Division of the Planes.*

Here the number of the planes of the fundamental figure is neither increased, nor is their figure changed, as
is the case with all the preceding alterations, but each
plane is divided into a greater or lesser number of smaller
planes that meet together under very obtuse angles.

The number of compartments into which a plane is
divided, is two, three, four, and six.

The dividing edges run either parallel to the diagonal,
or from the centre of the plane of the fundamental figure
towards the angles, or towards the middle of the central
or terminal angles. Of the first we have an example in
the dodecahedral garnet; and of the second in grey copper-ore and diamond.

v. *Multiplied*

v. *Multiplied Alterations.*

The various alterations of the fundamental figures just enumerated, occur singly or several together in the same fundamental figure. In the latter case, they are placed either beside each other, when they are said to be *co-ordinate*, or on one another, when they are said to be *superimposed.* The alterations are considered to be co-ordinate, when they occur in *different places* of the same fundamental figure; of this we have an example in fluor-spar, when the cube is bevelled on the edges, and truncated on the angles. They are named *superimposed,* when they occur in the *same part* of the fundamental figure, and when the first alteration is modified by a second, as in a prism which is bevelled on the terminal planes, and truncated on the bevelling edges. Sometimes, as in topaz, three or more superimposed alterations occurring together in the same figure. Crystallisations frequently occur which are so modified, that they may be described in different ways, and referred sometimes to one, sometimes to another fundamental figure. This gives rise to two modes of description, viz. the *representative* and the *derivative.* If a crystal is described as it appears to the eye at first view, without any reference to its relation to other crystallisations of the same mineral, it is said to be described representatively. But if in the description we attend to its relations with the other crystals of the same mineral, and also to its derivation from these, it is described derivatively. Thus, in calcareous-spar, we meet with forms, which, if described derivatively, would be considered as very low six-sided prisms, acuminated

on

on both extremities with three planes, the planes set on
the alternate lateral planes; and the summits of the acuminations so deeply truncated, that they touch the unaltered lateral planes in a line. But on a first view this
figure presents nothing prismatic; and if ignorant of its
origin from the prism already mentioned, we would rather consider it as a flat, double, three-sided pyramid, in
which the lateral planes of the one are set on the lateral
planes of the other, and the summits and the angles on
the common basis deeply truncated. In the same manner, many very broad prisms, as in rock-crystal, at
first sight appear like tables, but must be considered
as prisms, on account of their derivation and other relations.

The derivative mode is the most interesting and useful, and is that which ought to be followed whenever it
is possible.

In those cases, however, where the choice of the fundamental figure is optional, and when it is not determined
by tracing it from other crystallisations, we give the preference to that figure which enables us to describe the
crystal with the greatest facility and accuracy, and in
the shortest manner. It is sometimes advantageous, and
also facilitates our conception of the crystal, when we
unite together in our description both the modes, using
the derivative as the principal one. Thus many varieties of the cube and the rhomboid are more clearly expressed, when we describe them as double three-sided
pyramids, in which the lateral planes of the one are set
on the lateral edges of the other.

The different modes of describing crystals, depend on
the transitions that so often occur between them, by which
one figure, owing to a succession of modifications, gradually

dually passes into the other. Thus the cube, by the truncation of its angles, passes into the perfect octahe-dron. At first the truncating planes on the angles of the cube are small, but become gradually larger and larger until they touch each other, when the crystal exhibits a form intermediate between that of the cube and the oc-tahedron. If the truncating planes still increase in size, they become larger than those of the cube, and are now the principal planes of the figure, while those of the cube are alterating planes, and the whole represents an·octa-hedron truncated on the angles. If the original planes of the cube, which now form truncating planes in the angles of the octahedron, become smaller and smaller, and at length entirely disappear, the perfect octahedron is produced.

The modifications that give rise to these transitions are the following.

1. *Alterations taking place in the proportional magnitude of the planes between themselves.*

   Some planes increase in size, while others diminish, and thus one figure is changed into another. When the alternate lateral planes of the octahedron become larger, while the others diminish, a tetrahedron is formed, or the octahedron passes into the tetrahe-dron.

2. *Alterations in the angles under which the planes meet.*

   Thus the common dodecahedron, by the increasing obtuseness of its angles, at length passes into the cube.

3. The *convexity* of the faces of the crystals, which is sometimes occasioned by the division of the planes.

4. *By*

4. *By the newer or alterating planes becoming gradually larger at the expence of the original planes, which are at length totally obliterated.*

These changes are produced either by truncation, bevelment, or acumination : the transition of the cube into the octahedron, is an example of the first : the transition of the octahedron into the icosahedron, by the bevelment of the angles of the octahedron, of the second ; and the third is exemplified by the transition of the tetrahedron into the rhomboidal or garnet-dodecahedron by the acumination of each of the angles of the tetrahedron by three planes.

5. *By the aggregation of crystals.* Thus six-sided tables heaped on one another form six-sided prisms.

All the crystals that lie between two principal crystals, and form the transition of the one into the other, constitute what is called a *transition-suite.* These vary in extent, and sometimes they form circles, so that the last member of the suite passes into the first, or they form a straight line, and diverge into numerous branches.

Those mineral species that occur crystallised, are generally characterised by a particular suite of crystals, which does not occur in the other species. There are, however, mineral species very different from each other in their external characters, in which we meet with the same suite of crystals ; and still more frequently do we meet with species that exhibit not the whole suite of crystals of another species, but a greater or smaller portion of it. Thus there is an extensive suite of crystals which extends from the icosahedron, through the dodecahedron,

Q   the

the cube, and the octahedron, into the tetrahedron, part of which we sometimes meet with in mineral species, but never the whole, only a larger or smaller portion of it.

That member of the suite of crystals of a mineral species, from which all the others originate or proceed, is named the *Fundamental Crystallisation*. In those suites of crystals which form circles, it is often optional which of the figures we assume as the fundamental one ; and in those which are disposed in lines, it is sometimes of little importance at which end we begin our description. Still we always select that crystallisation which occurs the most frequently, and the most distinct in the species, and derive all the others from it.

In mineral species the crystals never appear isolated, but form a kind of progression, and pass gradually into each other. It follows from this important and highly interesting fact, that when a few crystals of a species are known, probably all the intermediate members of the series, which can be easily pointed out by crystallo-graphy, and which have not been found, may be expected to exist in nature, because the cause which produced the one part of the series may also have formed the others.

C. *The Magnitude of Crystals.*

This character is useful, not only in the description of varieties, but also in that of species, because in each mineral species the crystals appear to have a determinate range of magnitude. We have here to attend to the *absolute magnitude,* and also to the *relative magnitude* of crystals.

Crystals

Crystals in regard to their *Absolute Magnitude*, are divided into

α. *Uncommonly large*, when the crystal is two feet and upwards in length. The expression intimates that it is rare. Example, rock-crystal.

β. *Very large*, from two feet to six inches in length. Examples, rock-crystal, quartz, beryl, calcareous spar, and felspar.

γ *Large*, from six inches to two inches in length. It is a very frequent size. Examples, leadglance, garnet, rock-crystal, &c.

δ. *Middle-sized*, from two inches to half an inch. Examples of this magnitude are common, we shall only mention galena or lead-glance, ironpyrites, fluor-spar, calcareous-spar, and garnet.

ε. *Small*, from half an inch to the eighth of an inch. Examples, fluor-spar, calcareous-spar, &c.

ζ. *Very small*, from the eighth of an inch in length, until it is so minute as scarcely to be visible to the naked eye. Examples, native silver, grey copper-ore, spinel, &c.

η. *Microscopic*. When crystallised, but the form no longer distinguishable by the naked eye. Examples, gold, galena or lead-glance, &c.

In determining the *Relative Magnitude* of crystals, we use the following terms.

α. In the prism.

aa. For the length.
Short or *low*.
Long or *high*.

bb. For

bb. For the breadth and thickness.

Broad, when the breadth is greater than the thickness.

Acicular or *needle-shaped,* when the prisms are so thin, that the planes are seen with difficulty.

Capillary, when the planes of the crystals are no longer visible.

β. In the pyramid.

aa. For the length.

Short or *low.*

Long or *high.*

bb. For breadth and thickness.

Broad.

Lance-shaped, allied to acicular.

γ. In the table.

aa. For the length and breadth.

Longish, when one dimension of the lateral planes is greater than the others.

bb. For the thickness.

Thick and *thin.*

ẟ. Crystals in which all the dimensions are nearly alike, are named *tessular.*

D. The

D. *The Attachment of Crystals.*

WERNER understands by *attachment*, the connection of single crystals with massive minerals, and the aggregation of crystals together. According to the tabular view, the first distinction is into

A. *Solitary*, and this again into *loose*, *imbedded*, and *superimposed*.

    α. *Loose.* Crystals are said to be loose, when they are not connected with any other mineral.

    β. *Imbedded.* Crystals are said to be imbedded, when they are completely inclosed in another mineral. They are crystallised on all sides, or are said to be all around crystallised, and must therefore have been formed at the same time with the mineral in which they are imbedded. We cannot conceive them to have been of anterior origin to the basis in which they are contained ; for, on this supposition, we must conceive them to have remained suspended in space until the basis was formed around them. Nor can we admit them to be of posterior origin, because the crystals have impressed their form on the basis, and portions of the basis are sometimes contained in the crystals, and the crystals at their lines of junction, occasionally intermixed with the basis. Examples of this we have in garnets, imbedded in serpentine, or garnets in mica-slate.

    γ. *Superimposed.* When crystals rest upon the surface of another mineral, and are firmly attached

to

to it, they are said to be superimposed. No regular planes occur at their point of attachment, on the contrary, they take the impression of the kind of surface on which they rest. Hence it would appear, that probably they are often of posterior origin to the basis on which they rest.

The second distinction is into

B. *Aggregated.* Here there are two distinctions.

   a. Where a determinate number of *crystals grow together in a determinate manner, and these differ,*

     1. *With respect to number.*

       i. *Pair-wise* (twin crystals).

       ii. *Three together* (triple crystals).

       iii. *Four together,* (quadruple crystals).

     2. *With regard to the manner of their intersection,*

       i. *Penetrating one another.*

       ii. *Intersecting one another.*

       iii. *Adhering to one another.*

   Twin crystals are formed by two crystals penetrating, intersecting, or adhering to one another. Of the *first* we have an example in felspar, where they penetrate one another in the direction of their thickness; in gypsum, where they penetrate one another in the direction of their breadth; and in calcareous-spar, where they penetrate one another in the direction of their length. Of the *second* we have an example in cross-stone, where

where the crystals intersect each other, and form a kind of cross, and have a common axis ; and of the *third* in spinel, where the crystals adhere only by some of their planes. *Triple crystals*, occur in spinel and calcareous-spar.

Quadruple crystals, occur rarely, as in tin-stone.

b. *Where there are many crystals together, but merely simply aggregated;* and these are either, 1. *On one another;* 2. *Side by side,* or adhering laterally to one another ; and, 3. *Promiscuous.*

The *first* occurs principally in tessular crystals, as in galena or lead-glance, and calcareous-spar. The *second* occurs in amethyst, where the pyramids or prisms are parallel among themselves. The *third* occurs principally in long and broad figures, as in tables and prisms. We have examples of it in grey ore of antimony, where very long and nearly needle-shaped crystals cross one another in different directions ; also in tabular crystals, and this kind of tabular aggregation has much resemblance to the cellular external shape.

c. *Where there are many crystals together, but doubly aggregated.*

This kind of aggregation is distinguished from the foregoing by its forming groupes that exhibit shapes resembling bodies in common life.

i. *Scopiform* or *fascicular.* Is composed of a number of thin prismatic crystals, diverging from their point of attachment, and thus forming

forming a kind of fasciculus or bundle. Examples, calcareous-spar and zeolite.

ii. *Manipular* or *sheaf-like*. Consists of a number of crystals that diverge towards both
ends, and are narrower in the middle, thus
resembling a sheaf. It occurs in prismatic
and tabular crystals. Examples, zeolite,
calcareous-spar, and prehnite.

iii. *Columnar*. Consists of very long needle-
shaped prisms, many of which are connected together in the direction of their length ;
and these columnar groupes sometimes cross
one another in different directions. Example, columnar heavy-spar, and white lead-
ore.

iv. *Pyramidal*, is composed of many long prismatic crystals that are parallel to one another, but of which those in the middle are
the highest, and the others decline on all
sides, from the central one. Example, calcareous-spar.

v. *Bud-like*, is composed of low, (generally) six-
sided pyramids, one of which is usually situated in the middle, and is surrounded by
a number of others, whose extremities are
directed towards one another. Here also
many groupes occur together. Example,
quartz.

vi. *Amygdaloidal*, is formed by tables disposed
around each other, in such a manner as to
form an amygdaloidal shape. Example,
straight-lamellar heavy-spar.

vii.

vii. *Rose-like*, is composed of very thin six-sided tables, which are repeatedly curved, and so connected together that it resembles a blown rose. It occurs in the variety of calcareous-spar called *rose-spar*, from Joachimsthal.

viii. *Globular*. Is composed of tables or cubes aggregated into a globular shape. Examples, iron-pyrites, and curved lamellar heavy-spar.

ix. *In rows*. When many crystals are superimposed on each other, in a straight direction, like the pearls in a necklace, they are said to be aggregated in rows. The flat three-sided pyramids of calcareous-spar, and the octahedrons of silver-glance or sulphureted silver-ore, afford examples of this kind of aggregation.

x. *Scalarwise*, in which many tessular crystals are arranged like steps of a stair. Example, cubes of corneous silver-ore.

HAUY'S CRYSTALLOGRAPHY.

Having given a short view of the Wernerian Crystallo-graphy, in the preceding pages, we shall now add an exposition of the method followed by HAUY in determining and describing the various regular forms that occur in the mineral kingdom.

R   We

We shall consider it in the following order.

1. Of the primitive forms.
2. Of the integrant molecules.
3. Of the laws of decrement.
    a. Decrements on the edges.
    b. Decrements on the angles.
    c. Mixed decrements.
    d. Intermediate decrements.
    e. Compound secondary forms.
    f. Of secondary forms, when the molecules dif-
       fer from parallelopipeds.
4. Difference between structure and decrement.
5. Of those crystals in which the half is turned round,
    and of others that intersect each other.
6. Of the symbols used to denote the particular laws
    of decrement which produce the secondary
    forms.
7. *Lastly,* Of the nomenclature of crystals.

1. *Of the Primitive Forms.*

Crystals in which there are several cleavages, can, in general, be split in the direction of these cleavages. The fracture-surfaces or planes thus exposed, are either parallel with all the planes of the crystal, or they are not. When parallel with the planes, the crystal retains its form, however much we split it. in the direction of its planes ; it only changes in magnitude, becoming smaller and smaller as we continue the splitting. On the contrary, when the cleavages are not parallel with any of the planes of the crystal, or but with few of them, and when we continue the division until they meet
together,

together, we obtain a regular figure, whose contour
is very different from that which the crystal first exhibited, and which may now be further subdivided in
the direction of its planes. without undergoing any further change of form. This new regular form is by HAUY
named the *Primitive nucleus ;* and the crystal whose form
is the same, the *Primitive form.*

Thus, if we split a cube of galena or lead-glance in the
direction of its cleavages, all the divisions will be parallel with the faces of the cube, and by continued
splitting in the same direction, the figure remains unchanged, but the size gradually diminishes. On the
contrary, if we begin to split a cube of fluor-spar in the
direction of its cleavages, it first loses its eight angles,
and we observe in their stead eight new shining triangular planes. By continuing the splitting, these new planes
become larger and larger, and the form of the crystal is
more and more changed ; at length, the six four sided
faces of the cube disappear, and the eight triangular
planes meet, and form a regular octahedron. If the splitting is still continued, the octahedron diminishes in magnitude, but does not change in form. The figure ABCDEFG
Fig. 80. Pl. VII. represents a cube of fluor-spar. If we
attempt to split it by sections parallel to its planes or
faces, we will obtain an irregular, not a foliated surface ;
but if a section is made in the direction of the line $g f$, parallel to the diagonal line A C, upon one of its faces B D,
and at an angle of about $54\frac{3}{4}°$, it will split readily, and
the solid angle A g h f will be detached, and the new face
$g f h$ will be an equilateral triangle. Fig. 81. Pl. VII.
shews the position of the cleavages on the eight angles of
the cube, which are marked with the letters a b c d $e f g$ h.
If we continue to split the cube in the directions of these

cleavages,

cleavages, we will first obtain equilateral triangular faces; and when the sections intersect each other, the equilateral triangular faces will disappear and become hexagons. When the original faces of the cube entirely disappear, an octahedron, with triangular faces, will make its appearance, which is represented in Fig. 82. Pl. VII. as the nucleus contained in the cube. According to HAUY, the octahedron is the primitive form, and the cube a secondary form, of fluor-spar ; whereas the cube of galena agrees with the primitive form, and also with the primitive nucleus of that mineral.

If we take a regular six-sided prism of calcareous-spar, Pl. III. Fig 1, and attempt to divide it parallel to the edge of the base, we will find three of these edges taken alternately in the upper base; for example, the edges lf, cd, bm, will admit of this division, while the other three of them will not. To succeed in the lower base, we must not make choice of the edges $l'f'$, $c'd'$, $b'm'$, which correspond with the upper edges ; but the alternate edges df', $b'c'$, $l'm$, Fig. 2 Pl. III. These six cuts will expose to view as many trapeziums. Three of these are represented in Fig. 2. Pl. III. ; namely, the two which come in the place of the edges lf, cd, and which are marked by the dotted lines $ppoo$; $uakk$, and that which comes in place of the lower edge $d'f'$, and which is marked by the dotted lines $nnii$. Each of these trapeziums will have a polish and lustre, from which it will be easy to perceive that they coincide with the natural joints or cleavage of the prism. We will attempt in vain to divide the prism in any other direction ; but if we continue the division parallel to the first cuts, it is obvious that the size of the bases will gradually diminish, while the prism itself will continue to grow shorter. Just when the bases disappear altogether, the

prism

prism will be converted into a dodecahedron, Fig. 3.
Pl. III. with pentagonal faces ; six of which, as *o o i* O *e*,
o I *k i i*, are the remains of the faces of the prism, and
and the six others EAI*oo*, OA'K*ii*, are the result of the
mechanical division.

If we continue the splitting, the faces at the ends will
preserve their figure and size, while the lateral faces will
continually diminish in length, till at last the points *o,k*
of the pentagon *o* I *k i i*, being confounded with the points
i, i, and the same thing happening with all the other
points similarly situated, each pentagon is converted into
a simple triangle, as we see in Fig 4. Pl. III. New
slices taken off, make the triangles disappear, so that
no vestige of the original prism remains. Thus we obtain the prismatic nucleus, or what on a general view
may be called the Primitive Form, (Fig. 5. Pl. III.)
which consists of an obtuse rhomboid *, the inclination
of whose planes is 105° 5′, and 75° 55′ and the plane
angles EAI and AIO 101° 32′ 13″, and 78° 8′.

In some minerals, however, as in those with a concealed foliated fracture, in quartz for example, we can seldom divide the crystal in the direction of the cleavage
until it is heated red-hot, and thrown into cold water ;
by this process rents take place in the directions of the
cleavage, and enable us to s lit it regularly.

In other minerals, where this practice is inappli-
cable, we judge of the cleavage by the reflection of light
sometimes observable in their interior, when they are
turned in different directions before a strong light.
Lastly,

---

* By a *rhomboid*, HAUY means a figure rounded by six equal rhombuses,
parallel two and two.

Lastly, in those cases where no cleavage is discoverable, Haüy determines its position, and the figure of the primitive nucleus, and primitive form, by conjecture from the appearances presented by the forms of the secondary crystals.

However varied the crystallisations of a mineral species may be, all of them, according to Haüy, have the same primitive form, and the planes of this form are always parallel, either with actually existing cleavages, or with the conjectural cleavages of which we consider it to be composed. On the contrary, in different mineral species, the primitive form may vary more or less. But all these different forms can be reduced to some one of the six following :

1. The parallelopiped. Fig. 8. Pl. III.
2. The octahedron. Fig. 9. Pl. III.
3. The tetrahedron. Fig. 10. Pl. III.
4. The regular six-sided prism. Fig. 11. Pl. III.
5. The dodecahedron with rhomboidal faces, equal and similar. Fig 12. Pl. III.
6. The triangular dodecahedron with triangular faces, consisting of two six-sided pyramids, in which the lateral planes of the one are set on the lateral planes of the other. It is also named bi-pyramidal dodecahedron. Fig. 13. Pl. III.

I. *Parallelopiped.*

This is a solid figure, bounded by six faces, parallel to each other, two and two. Thus, for example, a cube is a parallelopiped. It is evident, from this definition, that there may be a vast number of parallelopipeds, differing
from

from each other in the proportional length and breadth
of their faces, and in the angles which these faces make
with one another. About forty parallelopipeds have been
hitherto observed in the mineral kingdom. They may
be divided into nine different kinds.

1. *Cube*, is a well known rectangular figure, bounded
   by six square faces, all equal to each other.
   Fig. 14. Pl. III. Examples, common salt, galena or lead-glance, common iron-pyrites, native
   gold, silver, and copper.
2. *Right quadrangular prism with square base*, Fig. 15.
   Pl. III. It is a cube somewhat longer in one
   direction than in another; of course the only way
   in which it can vary, is in the relative lengths of
   two contiguous faces, the base, and one of the faces of the prism. For all the faces of the prism
   are, of necessity, equal and similar, being all
   rectangles. There are seven species which have
   this primitive form. In four of them the prism
   is shorter than the square base, while in three it
   is longer. Meionite, Wernerite, Epsom salt, and
   zeolite, have the prism shorter than the base;
   while in Vesuvian, red lead-ore, and titanite, it
   is longer.
3. *Right quadrangular prism*, but in which the *base*,
   instead of a square, is *a rectangle;* of course it
   admits of greater variation than the preceding,
   as both the length of the base, and of the prism,
   may vary. Fig. 16. Pl. III. Examples, eu-
   clase and prehnite.

4. *Right*

136　　EXTERNAL CHARACTERS

4. *Right quadrangular prism. in which the base is a rhomb*, Fig. 17. Pl. III. Examples, celéstine and. mica.

5. *Right quadrangular prism, in which the base is an oblique-angled parallelogram*, Fig. 18. Pl. III. Examples, epidote or pistacite, and axinite.

6. *Oblique-angled quadrangular prism, the base of which is a rhomb*, Fig. 19. Pl. III. Hornblende and augite are species which have this kind of primitive form.

7. *Oblique-angled quadrangular prism, the base of which is an oblique-angled parallelogram*, Fig. 20. Pl. III. Felspar is an example of a mineral with this kind of primitive form.

8. *Oblique-angled quadrangular prism, the base of which is a rectangle*, Fig. 21. Pl. III. Example, borate of soda or borax.

9. *Rhomboid with obtuse summits*, Fig. 22. Pl. III. In rhomboids of this kind there are two solid angles opposite each other, which differ from the other six. They are composed each of three obtuse plane angles, meeting together in a point; whereas the six others are formed of two acute and one obtuse angle. The line joining these obtuse solid angles is called the *axis* of the crystal, and the angles themselves constitute the *summits* of the crystal. The axis is the *shortest* line joining any two opposite angles in the respective rhomboids. Examples, calcareous-spar, chabasite and quartz.

10. *Rhomboid with acute summits*, Fig. 23. Pl. III. Two of the eight solid angles of this figure differ
fer

fer from the other six. They are formed by the
inclination of three acute plain angles to each
other, whereas the other six are composed of
two obtuse and one acute plain angle. The
line joining these two acute and opposite solid
angles, is called the *axis of the crystal*, and the
angles themselves are called the *summits*. The
axis is the *longest* line joining any two opposite
angles of the crystal. Examples, corundum,
iron glance, and vitriol of iron.

II. *Octahedron.*

This may be defined a double four-sided pyramid, in
which the lateral planes of the one are set on the lateral
planes of the other. There are four different kinds of
this form, viz.

1. *Regular octahedron*, Fig. 24. Pl. IV. In this figure the triangular faces are equilateral and
equiangular, and, of course, the base of the two
pyramids is *a square*. Examples, diamond, spinel, fluor-spar, and magnetic ironstone.

2. *Octahedron*, in which the pyramids have *rectangular bases*, Fig. 25. Pl. IV. Each triangular face,
of course, is isosceles, the two angles at the base
being equal, and the angle at the summit different. It may be either acute, rectangular, or obtuse, according to the length of the rectangular
base of the pyramids. The faces are equal and
similar, four and four. Examples, white lead-
ore, and sulphat of lead or lead-vitriol.

3. *Octahedron*, in which the pyramids have *a square
base*, Fig. 26. Pl. IV. The pyramids, in gene-

S   ral,

ral, are very low, when compared with the size
of the base, though this is not always the case.
In anatase, for example, they are long, and of
course the solid angle at their summit is composed of very acute plane angles meeting at a point.
Examples, zircon, tinstone, and cross-stone.
4. *Octahedron,* in which the pyramids have *a rhomboidal base,* Fig. 27. Pl. IV. They vary from each
other in the height of the pyramids, and the
angles of the rhomb constituting the common
base of the pyramids. Examples, sulphur and
sphene.

III. *Regular Tetrahedron,*

Is a figure bounded by four equilateral and equiangular triangles ; or it may be described as a three-sided pyramid, terminated by a triangular base, Fig. 28. Pl. IV.
Examples, grey copper, and copper-pyrites or yellow
copper-ore.

IV. *Regular Six-sided Prism,*

Is a prism composed of six equal rectangles, and terminated at each extremity by a hexagonal plane or base,
Fig. 29. Pl. IV. They differ from each other in the
height of the prism compared with the diameter of the
base. Examples, emerald, apatite, and strontianite.

V. *Rhomboidal Dodecahedron,*

Is a solid bounded by twelve equal rhombs, Fig. 30.
Pl. IV. Example, garnet.

VI.

VI. *Triangular Dodecahedron.*

It consists of two six-sided pyramids, joined base to base, and the common base, of course, is a regular hexagon. Example, witherite, Fig. 31. Pl. IV.

Of all these primitive forms, by far the most frequent are the parallelopiped and the octahedron. The six-sided prism is also pretty frequent, but the other three primitive forms, viz. the tetrahedron, the rhomboidal dodecahedron, and the triangular dodecahedron, occur but rarely.

2. *Of the Integrant Molecules.*

Sometimes the primitive nuclei, and primitive forms, besides their divisions or cleavages parallel with their planes, exhibit others in other directions, which are not parallel with their planes, or at least not with all of them, and thus give rise to a new figure, more simple than that obtained by the first mechanical division.

Thus, if we continue to cut slices from a six-sided prism, by cuts parallel to the lateral planes of the prism, we will divide the whole prism into a number of triangular prisms. This will be evident to the eye, by inspecting Fig. 32. Pl. IV., which represents the basis of a six-sided prism, divided into triangular prisms by such continued divisions. Sometimes a parallelopiped admits of divisions in other directions, besides those parallel to its faces. Suppose the rhomboid AA′KH, Fig. 33. Pl IV. divisible both in the direction parallel to the six rhombs which constitute its faces, and likewise in planes passing through

through the oblique diagonal AO, the axis A'A, and the edge A'O, comprehended between the diagonal and the axis, the consequence of such a division would be, that the rhomboid would be separated into six tetrahedrons.

These tetrahedrons are represented in the figure sur‑ rounding the original rhomboid ; and, to aid the concep‑ tion, the same letters are employed to denote the same parts in the rhomboid, and in the tetrahedrons into which it is conceived to be divided. These forms are what Hauy names *integrant molecules*, and which he conceives to be the form of the ultimate integrant atom of the mi‑ neral in question. Hauy has also found, that the inte‑ grant molecules of all crystals, supposing them capable of being discovered by mechanical division, may be redu‑ ced to three species, viz. the Tetrahedron, Fig. 34., Pl. IV. the Triangular Prism, Fig. 35. Pl. IV, and the Parallelo‑ piped, Fig. 36. Pl. IV. These three forms of the inte‑ grant molecules, like those of the primitive nuclei, and primitive forms, vary in their dimensions, and in the magnitude of their angles.

These integrant molecules arrange themselves in a va‑ riety of ways, and thus give rise to the different seconda‑ ry forms. Hauy shews that these secondary forms may be accounted for, by supposing that layers of integrant molecules, arranged so as to form plates, are applied suc‑ cessively to all the faces of the primitive crystal, while each successive plate diminishes in size by the abstraction of a determinate number of integrant molecules (or pa‑ rallelopipeds), either parallel to the edges, or the diago‑ nal of the faces, or in some other intermediate direction. Sometimes decrements take place at once on all the edges, as when the rhomboidal dodecahedron is formed from the cube ; or upon all the angles, as when the regular octa‑ hedron

hedron is formed from the cube. Sometimes they take
place only on certain edges, or certain angles. Sometimes they are uniform, so that only one law exists of
decrements, by one, two, three, &c. ranges, which acts upon different edged angles. Sometimes the law varies
from one edge to another, or from one angle to another;
and this happens chiefly when the nucleus has not what is
called a *symmetrical form*, as when it is a parallelopiped,
whose faces differ in the respective inclinations of their
faces, or in the measure of their angles. In certain cases,
the decrements on the edges concur with those in the
angles to produce the same crystalline form. It happens
likewise, sometimes, that the same edge, or the same
angle, undergoes different laws of decrement, which succeed each other. And, finally, there are a great many
cases, where the secondary crystal preserves faces parallel to those of the primitive form, and which combine
with the faces produced by the decrement to modify the
figure of the crystal. If, in the midst of such a diversity
of laws, sometimes acting solitarily, and sometimes in
combination, upon the same primitive form, the number
of ranges subtracted were likewise very variable; if, for
example, there were decrements of 20, 30, 40, or a greater
number of ranges of molecules, as is very possible in conception; the multitude of forms which might exist in
each mineral species, would be sufficient to confound the
imagination; and the study of crystallography would
present an immense labyrinth, from which, even when
assisted by the theory, it would be difficult to extricate
one's-self.

But the force which produces the subtractions, appears
to have a very limited action. Generally these subtractions take place only by one or two rows of molecules.

None

None have hitherto been observed beyond six rows. But such is the fecundity united with this simplicity, that, supposing we confine ourselves to decrements by one, two, three, and four rows, and exclude those that are mixed or intermediate, we find that the rhomboid is susceptible of 8,388,604 varieties of crystallisation. Doubtless many of these varieties do not exist in nature. But there is reason to expect, that discoveries in the field of inquiry will be made in great numbers for a long time to come.

We have already remarked, that besides the parallelopiped, there are two other shapes which the integrant molecules assume; namely, the tetrahedron, and the triangular prism.

It is worthy of notice, that the tetrahedral and prismatic molecules are always arranged in such a manner in the interior of primitive and secondary crystals, that, taking them in groups of 2, 4, 6, or 8, they compose parallelopipeds; so that the ranges subtracted by the effect of decrement, are nothing else than these parallelopipeds. These parallelopipeds are by Hauy named *Subtractive Molecules.* They are always substituted in place of tetrahedrons or triangular prisms, in considering the decrements which produce the secondary forms in these cases.

3. *Of the Laws of Decrement.*

In order to enable the reader to understand easily these various decrements, we shall give the following illustrations from Hauy *.

1. *De-*

---

* I have here used the statement of Hauy, as given in the article Crystallography, in the *Edinburgh Encyclopædia.*

a. *Decrements on the Edges.*

Let us suppose that the primitive form of a mineral species is the cube; but that secondary crystals of the same species likewise occur, having the form of the rhomboidal dodecahedron. How is this dodecahedron derived from the cube? Let us suppose, as may be done in every case, that the integrant molecule of this species is a cube; it follows that the primitive cubic crystal is formed by the congeries of a number of cubes. Suppose these cubes of such a size that an edge of the primitive crystal is composed of seventeen of these small cubes applied side by side. Of course every face of the primitive crystal will be composed of 289 squares, consisting of the bases of so many integrant molecules. According to this supposition, the primitive crystal will be a congeries of 4913 little cubes. Let us now suppose, that a square, consisting of the thickness of one integrant molecule, be applied to every face of the primitive crystal; but that, instead of being of the size of the face of that crystal, it be less than it by a single row of integrant molecules all round, so that its side, instead of 17 little cubes, contains only 15; and of course it contains only 225 little cubes, instead of the 289 that go to the formation of the face of the primitive crystal. Upon each of these first plates applied all round to every face, let another plate be applied similar to the first, but less than it by a row of integrant molecules, so that the side contains only 13 squares, and the whole plate only 169 squares. Let six other plates be applied in succession to each of the faces, diminishing by a row of little cubes all round, so that the sides of each consist of 11, 9, 7, 5, 3, I, squares, respectively. It is obvious, that, by this process.

cess, we have raised upon each of the six faces of the
cube a four-sided pyramid, the faces of which, instead
of being smooth, will, by their constant diminution in
bulk, represent the steps of stairs. These pyramids
having each four faces, constitute small 24 triangular faces; so that, by this process, we have converted
the cube into a new crystal. It would seem, at first
that this new crystal ought to have 24 triangular faces;
but a little consideration will satisfy us, that the two
adjacent triangular faces, in each pyramid, are in the
same plane, and form together a rhomb; so that, in
fact, the cube has been converted into a rhomboidal dodecahedron. Fig. 37. Pl. IV. represents the cubic nucleus,
with the pyramids raised upon three of its faces; and
Fig. 38. Pl. IV. represents the rhomboidal dodecahedron
formed in this manner. This is an example of a secondary
crystal formed by decrements on the edges of the plates.
Suppose us in possession of such a crystal, it is easy to
see how, by mechanical division, the cubic nucleus
might be extracted. We would have only to cut off
all the solid angles formed by four plain angles, by
slices parallel to the shorter diagonals EO, OI of the
rhombs.

In the preceding example, each plate was only of the
thickness of one integrant molecule, and the decrement
was only one row of integrant molecules all round; but
we might have supposed the thickness of the plates to
have equalled two or more integrant molecules, and the
decrements might have been equal to two rows of integrant molecules, or more, at once. In that case, the
form of the secondary crystal obtained would have been
different from the rhomboidal dodecahedron.

It

It will be necessary here to explain the meaning of two terms, which we will have occasion to employ frequently hereafter. *Decrement in breadth* is used when the thickness or height of the plate is only equal to one integrant molecule; but one, two, three, &c. rows of molecules all round, we conceive to be abstracted from the breadth of each succeeding plate. *Decrement in height* is used when the plates only diminish by one row of integrant molecules in breadth, but their height may be equal to two, three, &c. molecules. In such cases, the decrement is expressed by saying, that it takes place by two, three, &c. rows in height.

It will be worth while to give another example of a secondary crystal formed by decrements on the edges of the faces. The primitive form of iron-pyrites is a cube; but, among a great variety of secondary crystals, there is one which occurs in the form of a dodecahedron with pentagonal faces. This crystal is represented in Fig. 39. Pl. IV. where the cubic nucleus may likewise be seen. From the inspection of that figure, it will be obvious, that, instead of a four-sided pyramid, as in the former case, a kind of wedge is formed upon each face of the cubic nucleus, which may be conceived to be the pyramid elongated in one direction. This wedge upon one of the faces of the cube, is represented by OO' ι n I I'. In this case, the decrements may be conceived to take place by two ranges in breadth between the edges OI and AE, II' and OO', EO and E'O'; and in the same manner upon the opposite faces; while, at the same time, they take place by two ranges in height between the edges EO and AI, OI and O'I', OO' and EE'. We see that these decrements take place upon the different faces of the cube in three directions, which cross each other at right angles. The decrement, by two

T   ranges

ranges in breadth, tending to produce a face more inclined than that which results from a decrement by two
ranges in height ; the consequence must be, that the
structure of plates does not terminate in a point, as in
the first example, but in a wedge. The lines $p\,q$, $t\,n$,
Fig. 39. Pl. I V. represent the summits of two of these
wedges. If we compare these summits $p\,q$, $t\,n$, with the
summit $r\,s$ of the wedge which covers the face EOO'E'
of the cubic nucleus, it will be easy to perceive that these
three lines are perpendicular to each other respectively.
Fig. 40. Pl. IV. represents the cubic nucleus with wedges
raised upon two of its contiguous faces by means of plates
pursuing decrements according to the law above described. The same letters are applied to the same parts of
the crystal in Figs. 39. and 40. Pl. IV. At S' is seen
the extremity of the summit of a third wedge raised upon
a third face of the cube. Each trapezium, such as O $p\,q$ I
Figs. 39. and 40. Pl. IV. being in the same plane with
the triangle O t I belonging to the adjacent wedge,
both together conspire to form the pentagon p O t I q, so
that the secondary crystal formed by these decrements,
instead of 24 faces, has only 12 pentagonal faces, and is
therefore a dodecahedron as well as the first example, but
a dodecahedron of a different kind.

We shall give a third example of these kind of decrements, because it contains something peculiar in it, but
which often takes place in the formation of secondary
crystals ; and it is requisite that the reader should be
aware of it. The dodecahedron represented in Fig. 6.
Pl. III is a secondary crystal of calcareous-spar. In
it the edges EO, OI, IK, &c. where the two opposite
pyramids join, coincide with the edges of the primitive
                                                nucleus,

nucleus, as may be perceived by inspecting Fig. 7.
Pl. III. The decrements set out from these edges, and
do not take place at all upon the other six edges of the
principal nucleus EA, AI, AG, OA′, &c. Now, it is
easy to conceive, that the edges of the plates laid upon
the primitive nucleus, form, by their sum, as many triangles E s O, I s′ O, E s′ O, &c. resting upon the edges
from which they set out; and as these lines are six in
number, there will be 12 triangles, six above, and as
many below; and all these triangles will be scalene,
in consequence of the obliquity of the edges from which
the decrements set out.

With respect to the other edges of the plates of superposition, they will be so far from experiencing any decrement, that they will, on the contrary, augment, because
they must always remain contiguous to the axis of the
crystal, just as happens when the primitive crystal increases in size by the superposition of new plates, without
undergoing any change of form. It is the province of
mathematics, combined with observation, to determine
the law of decrement upon which this dodecahedral form
depends. If we suppose a decrement of one range, it
may be demonstrated that the two faces produced on each
side of the edge from which the decrement set out, will
be in the same plane, and parallel to the axis of the primitive crystal, circumstances which do not suit the present case. If we suppose a decrement of two ranges in
breadth, it may be demonstrated that the result will be a
dodecahedron similar to the one which we are considering. Fig. 41. Pl. V. represents one of the pyramids of this dodecahedron formed by the superposition
of plates following the law of decrements by two ranges

of

of particles. The line E *s* represents an edge of this pyramid such as it appears to the eye, E *s* such as it really exists; but the distance *s s'* is not sensible, in consequence of the extreme minuteness of the size of the intermolecules, by the abstraction of which the pyramids are formed. The same reason prevents the channels or steps of stairs upon the pyramids from being sensible. Though in some cases, when secondary crystals are formed with great rapidity, these channels may be perceived by the naked eye.

2. *Decrements on the Angles.*

Decrements on the edges, which have been just described, are not sufficient to account for all the diversity of forms which secondary crystals assume. To give an example; mineral species, the primitive form of whose crystals is the cube, are found crystallised in secondary forms, some of which are rhomboidal dodecahedrons. and others regular octahedrons. The formation of the rhomboidal dodecahedron has been explained above, by means of decrements on the edges. At first sight, it would appear that the octahedron might also be derived from the cube by decrements on the edges. We have only to take two opposite faces of the cube, and to suppose a four-sided pyramid raised upon each by means of decrements on the edges of the plates successively applied. While this is going on upon these two faces, we may suppose that the other four faces of the cube remain unaltered. Each of these two pyramids may be supposed to prolong itself downwards till they meet. The consequence would be

an

an octahedron enveloping the cubic nucleus; but it may
be demonstrated, that no law of decrement whatever
could in this case form an octahedron with equilateral
triangular faces, which is the case with the octahedron
derived from the cube. Besides, if we have recourse
to mechanical division, in order to obtain the cubic nucleus from this kind of octahedron, we shall find that the
solid angles of the cube coincide with the central points of
the eight faces of the octahedron, which could not be the
case if the octahedron had been formed in the way we
have been supposing. But if we suppose the decrements
to take place parallel to the diagonal of the faces of the
cube, all difficulty vanishes; we obtain the regular octahedron without difficulty. Such decrements are called
decrements on the angles.

Let OI I'O' Fig. 42. Pl. V. be one of the faces of the
cubic nucleus, divided into a number of little squares,
which are the bases of as many molecules. We may conceive these molecules arranged in two different ways; they
may be parallel to the edges, as is the case with the molecules, *a, n, q, r, s', t', v', z', s'*; or they may be arranged in the direction of the diagonals, as is the case with
the molecules *a, b, c, d, e, f, g, h, i,* and likewise with
the molecules *n, t, l, m, p, o, r, s,* and likewise with the
molecules *q, v, k, u, x, y, z.* One of these rows of molecules is represented separately in Fig. 43. Pl. V.

The molecules parallel to the edges of the square touch
by one of their faces, and the ranges themselves are simply placed contiguous to each other. The molecules parallel to the diagonals touch only by an angle, and the
ranges are indented into each other. When secondary
crystals are formed by this last kind of decrement, the
new faces are not merely channelled, as happens in the

case

case of decrements on the edges ; they are all bristled
with points, which being exceedingly minute, and all in
the same plane, escape the eye, so that the faces appear
smooth.

Having thus explained the meaning of the terms, let
us illustrate this kind of decrement by an example ; and
we cannot get a better than the formation of a regular
octahedron from a cubic nucleus. This is the consequence of the superposition of plates upon each face of
the cube with decrements of a single range of molecules
on the angles. Let AEOI, Fig. 44. Pl. V. A, be one of the
faces of the cubic nucleus subdivided into 81 little
squares, which are the bases of so many molecules, of
which the face is conceived to be composed. Fig. 44. B,
represents the first plate of superposition, which ought to
be placed above AEOI, Fig 44. A, in such a manner,
that the point e' corresponds with the point e ; the point
a' with the point a ; the point o with the point o ; and
the point i' with the point i. It is obvious, from this
manner of placing it, that the squares E e, A a, I i, O o,
Fig. 44. A, remain uncovered ; which is the initial effect of the decrement on the angles. We see likewise,
that the edges QV, PN, LC, FG, Fig. 44. B, exceed by
a range of molecules the edges EA, EO, OI, IA,
Fig. 44. A. This is necessary to prevent re-entering
angles, and is merely the consequence of the increase of
size of the crystal, without any change of form in these
quarters.

The upper face of the second plate of superposition, is
represented by BKHD Fig. 44. C. It must be applied to the first plate in such a manner, that the points e'',
a'', i'', o'', coincide with the points e', a', i', o', Fig. 44.
B, which leaves bare another row of molecules parallel

                                                            to

to the diagonal. This plate also increases by a row of
molecules at all its edges B, K, H, D, for the same reason as the first plate did.

The figure of these plates of superposition, which at
first was an octagon, has now become a square. It is no
longer necessary to continue the addition of rows of molecules at the edges ; so that the succeeding plates retain
the square shape, but constantly diminish in size, in consequence of the abstraction of a row of molecules from
each edge parallel to the diagonal of the face of the cubic nucleus. These different plates are represented by
Fig. 44. D, E, F, G, H, and I, in each of which the small
accented letters denote the points of the plate that coincide with the same letters in the preceding plate. Eight
plates are necessary, as appears from the Figure, and the
last of them consists only of a single molecule.

If we suppose the same number of plates, of the same
form, to be applied successively upon each face of the cubic nucleus, it is obvious that we raise upon each of the
six faces of the cube a four-sided pyramid. Hence it
would appear, at first sight, that the secondary crystal
would have 24 faces. Each of these faces will have four
edges, as must appear evident upon a little consideration,
and will have the form represented in Fig. 45. Pl. V. in
which the angle o is conceived to coincide with the angle
O of the cubic nucleus, and the diagonal $t x$ represents
the edge HK Fig. 44. C of the plate BKHD. The triangle $t o x$, being composed of those plates of superposition the edges of which undergo an increment, will be
much shorter than the triangle $t s x$ formed of those plates
of superposition whose edges undergo no increment ; because the number of the first is much smaller than that
of the second, they being to each other as 2 to 6.

Thus

Thus the surface of the secondary crystal is composed of 24 quadrilateral faces, arranged, three and three, round each angle of the cubic nucleus. But as in the decrements, by one range of molecules on the edges, the faces produced on both sides of the same edges are in the same plane, so in decrements by one range of molecules on the angles, the faces formed on the three sides of each angle are in the same plane. This plane is represented in Fig. 46. Pl. V. where the three quadrilaterals surrounding the angle of the cube o, coincide to form the equilateral triangle $m\ n\ s$. Thus the faces of the secondary crystal are reduced to eight equilateral triangles, and of course the figure is that of the regular octahedron.

If these decrements were to stop before they terminated in a point, the consequence would be, that faces would remain parallel to the original faces of the cube, and that the crystal would have fourteen faces, eight those of the octahedron, and six those of the cube ; so that it would at once have the form of the cube and of the octahedron. Nothing is more common than to find such crystals both in iron-pyrites and in galena or lead-glance.

If the decrements were more rapid, as, for example. if two or more ranges of molecules were abstracted, then the three trapezoids $s\ t\ o\ x$, $m\ t\ o\ r$, $n\ r\ o\ x$, (Fig. 46.) formed round the same solid angle of the nucleus, would not be in the same plane, but would be inclined upon each other, and the secondary crystal would have 24 trapezoidal faces.

As another example of this kind of decrement, let us take the rhomboid, Fig. 47. Pl. V. which differs somewhat from a cube by having acute angles. Let us suppose that the plates applied upon all the faces of this rhomboid suffer
decrements

decrements only at the angles contiguous to the summits A, O', and that these decrements take place by two ranges ; then, instead of 24 faces, only six would be form-ed ; and if we conceive these prolonged till they meet each other, they would compose a very obtuse rhomboid, which would be the secondary crystal. Fig.48.Pl.V. repre-sents such a rhomboid, with its primitive nucleus inclo-sed. We see that its summits, A, O' coincide with the summits of the primitive rhomboid, from which the de-crements commenced, and that each of its faces, as A e o i, corresponds with one of the faces AEOI of the nucleus, so that the diagonal which passes through the points e, i, is parallel to the diagonal EI of the face of the nucleus, and only somewhat more elevated. This kind of crystal is found among the secondary crystals of specular iron-ore or common iron-glance.

3. *Mixed Decrements.*

This name is applied to those decrements in which the number of ranges taken away in breadth and height give ratios, the two terms of which surpass unity. As, for example, decrements by two ranges of molecules in breadth, and three in height, or by three ranges in breadth and two in height, &c. It is easy to see that the theory may be with facility reduced to that of decrements, in which there is only one row of molecules taken away in one of the two directions.

4. *Intermediate Decrements.*

We have seen, that in the case of a decrement by one row of molecules round the same solid angle, the three faces produced are always in the same plane, and that, in

U that

that case, it is only necessary to consider the effect of the decrement with respect to one of the plane angles which concur to the formation of the solid angle, conceiving this effect to be prolonged over the neighbouring faces. In that case, the decrements on these last faces are considered as *subsidiary*, to favour the action of the principal decrement.

In general, whenever the solid angle of a primitive crystal undergoes decrements which tend to produce a face in its place, whatever the law may be to which we reduce the production of that face, there are always auxiliary decrements, the concurrence of which is necessary, in order that the new face may be of the requisite magnitude. Now, when the decrement which we consider in preference takes place, by two ranges of molecules, or by a greater number, the auxiliary decrements in continuity with it follow a particular law, which it is necessary to explain.

Let AA Fig. 49. Pl. V. be a parallelopiped of any kind which undergoes a decrement by two ranges on the angle EOI of its base AEOI. It is obvious that the edges of the plates of superposition will have the directions $b\,c$, $r\,s$, parallel to the diagonal EI, and so situated that there will be upon the sides OE, OI two rows of molecules comprehended between the angle O and the line $b\,c$, and likewise between $b\,c$ and $r\,s$. But, as has been already said, the plates applied upon the adjacent faces IOAK, EOA′H undergo likewise auxiliary decrements, which continue the effect of the decrement upon the angle EOI. But such, in this case, are the effects of these decrements, that the edges of the plates applied upon IOAK have the directions $c\,g$, $s\,t$; and those of the plates applied upon EOA′H the directions $b\,g$, $r\,t$. For
since

since the lower edge of the first plate applied upon
AEOI coincides with *b c*, and the height of this plate
corresponds to that of a single molecule, a little attention
will satisfy us that the plane *b c g*, which on one part
coincides likewise with *b c*, and on the other separates
from the base AEOI, by a quantity measured by O *g* the
height of a single molecule, is necessarily parallel to the
face produced by the decrement. The same holds with
the plane *r t s*. From this it follows, that if we suppress
the part situated above *r t s*, we will have a solid on
which the face *r t s* will represent the effect of decrement
that we are considering.

Now the directions *cg*, *st* of the plates applied upon the
face IOA K (and the same may be said of the face
EOA H) in consequence of the auxiliary decrements, are
neither parallel to the edge, nor to the diagonal of the
face, but intermediate between the one and the other.
This want of parallelism will become still greater, if
we suppose the decrements upon the angle of the base
EOI to take place by 3, 4. &c ranges. This is the kind
of decrement to which the name of *intermediate* has been
given. It is obvious, that it may take place in an infinite number of different directions, according as it deviates more or less from its two limits, the parallelism with
the edge, and the diagonal of the face

In cases similar to those of Fig. 49. Pl. V. we avoid
the complication introduced by these intermediate decrements, by supposing them comprehended under the principal decrement. But certain crystals exist, in which all
the three decrements round the same solid angle are intermediate. In such a case, the simplest of the three is
chosen as the principal decrement, and the other two considered

sidered as auxiliary. Fig. 50. Pl. V. represents a case of this kind ; cn, which is the edge of the first of the plates applied upon AEOI, is so situated, that on the side of OI there are three molecules subtracted ; while on the side OE there is only one : np, which is the edge of the first plate applied upon IOA′K, indicates three molecules subtracted from OI, and two from OA′ ; cp, which is the edge of the first plate applied upon EOA′H, shews the subtraction of two molecules on OA′, and only one on OE.

It is easy to see that the decrements take place relatively to the different faces situated round the angle O, as if the molecules that compose the different plates of superposition, being united invariably several together, compose other molecules of a higher order, and as if the subtraction took place by single ranges of these compound molecules. Thus there will be on the base AEOI a decrement of triple molecules by two ranges in height, since on one part the quadrilateral figure $cOnz$, which represents the base of a compound molecule, is equivalent to the bases of three simple molecules ; and, on the other, the line Op, which corresponds to the height of a plate of superposition, is equivalent to the height of two simple molecules. It is easy to conceive, likewise, that the decrement relative to the face EOA′H takes place by two ranges in height of double molecules, because $cOpx$ contains the bases of two simple molecules, and On is equal to the length of three simple molecules. In the decrement which takes place upon IOA′K there is a subtraction of one row of molecules triple in one direction, and double in the other.

Among these three decrements, the one which it appears most natural to adopt as the principal, is the second,

cond, which takes place upon the face EOA′H, because it is the one whose direction deviates the least from that of the diagonal EA′; or because it takes place by double molecules, which is a more simple decrement than the other two.

Suppose intermediate decrements on the two lateral angles G, G′ Fig. 51 Pl. V. of the face of a rhomboid, and that these decrements take place by ranges of double molecules, that is to say, parallel to the lines $u\,m$, $x\,y$, $u′m$, $x′y′$. It is evident that these decrements will produce above each rhomb of the primitive nucleus, such as SG $g″$G′, two faces which commencing at the angles G, G′, will converge towards each other, and come in contact in a line situated above the diagonal S$g″$, but inclined to that diagonal; so that the complete result of the decrement will be the formation of twelve faces disposed six and six towards each summit. Fig. 52. Pl. V. represents one of these solids, with its nucleus inscribed. It is a variety of calcareous-spar which sometimes occurs. The lines $a\,b\,a′$ shew the direction of a fracture parallel to the face G $g″$G′S of the primitive nucleus. It appears from this figure that the nucleus does not touch the secondary crystal, except by its lateral angles, which are situated in the edges BS′, D$s′$, C$s′$, &c. while in the dodecahedron of BERGMAN, represented in Fig. 6. and 7. and called by HAUY, *Metastic calcareous-spar*, the lateral edges of the nucleus coincide with those edges of the secondary crystal that constitute the common basis of the two pyramids, as is evident from inspecting Fig. 7. Pl. III.

Hitherto intermediate decrements have been observed only in a small number of instances, but they lead to forms as simple as the other, and give some curious results, which

which deserve to be studied in a mathematical point of view, without any reference to crystallography.

5. Compound Secondary Forms.

Simple secondary forms are those which proceed from a single law of decrement, the effect of which covers and conceals the nucleus, which only touches the surface of the secondary crystal by certain angles or edges. *Compound secondary forms* are those which are produced by several simultaneous laws of decrement, or by one law which has not reached its limit, so that faces remain parallel to the original faces of the nucleus, and which concur with the faces produced by decrement, to modify the form of the crystal. Suppose, for example, that the law which produces an octahedron from a cube (described above) should combine with that from which results the dodecahedron with pentagonal faces, Fig. 39. Pl. IV. The first of these laws would produce eight faces, which would have, for centres, the eight angles of the cubic nucleus. It is easy to see that each of these faces, that, for example, whose centre coincides with the solid angle O, Fig. 39. Pl. IV. will be parallel to the equilateral triangle whose sides pass through the points *p, s, t*. In like manner, the face whose centre coincides with the point O', will be parallel to the equilateral triangle, whose sides pass through the points *s, n, p'*. But the second law produces faces situated as the pentagons, cut by the sides of the triangles *p s t, s n p'*. Now the section of these triangles upon the pentagon *t O s O' n*, reduces the pentagon to an isosceles triangle, which has the line *t n* for the base, and the two other sides of which pass through the points *t, s*, and *n, s*. The same thing takes place with

with the other pentagons.   Hence it follows, that the se‑
condary crystal produced will be an icosahedron, bound‑
ed by eight equilateral triangles, and twelve isosceles tri‑
angles.

Fig. 53. Pl. V. represents this icosahedron, in which
the letters correspond with those of Fig. 39. Pl. IV. and
shew to the eye the relation between the two solids. But
this icosahedron has dimensions much greater than those
of the icosahedron which would be obtained by making
sections of the eight solid angles of the dodecahedron,
Fig. 39, Pl. IV. which are identified with those of the
nucleus.   This increase of size was necessary to preserve
the size of the nucleus.   This will be better understood
by the following illustration.

If we wished to obtain the nucleus from the icosahe‑
dron of Fig. 53. Pl. V. it is evident that the fractures
must be made in directions parallel to the edges r s, t n,
p q, Fig. 39. and 53, so that they should be equally
inclined upon the faces of which they form the junction.
These planes would pass at the same time through the
equilateral triangles p s t, s n p', &c. and we would obtain
the nucleus when they all met at the centres of the equi‑
lateral triangles.

It follows from this, that the nucleus, the edges of
which OI, OE, &c. Fig. 39. Pl. IV. were uncovered up‑
on the surface of the dodecahedron, is entirely enveloped
in the icosahedron, Fig. 53. Pl. V. excepting its solid
angles, which are only points, and which constitute the
centres of the equilateral triangles.   This being under‑
stood, in order to form an accurate idea of the structure
of the icosahedron, we must conceive that the plates ap‑
plied to the nucleus for a certain period undergo decre‑
ments only at the angles, as if the secondary solid were
to

to be a regular octahedron. Beyond this term (the decrement on the angles continuing always) a new decrement takes place and combines with the preceding ; and this new decrement being relative to the dodecahedron, produces the twelve isosceles triangles. In this manner we see how the nucleus is entirely inclosed in the dodecahedron, excepting the solid angles. The first plates of superposition, which only underwent a decrement on the angles, continued to envelope the nucleus by those portions of their edges which underwent no decrements. It is sometimes necessary to suppose, in this manner, different epochas to different decrements, which concur to produce a compound secondary form, when we wish to give a particular account of the mechanism of the structure.

From this statement it follows, that the distance between the centres of the equilateral triangles $p\,t\,s$, $q\,t\,s'$, Fig. 53. Pl. V. ought to be equal to the corresponding edge OI of the nucleus, Fig. 39. Pl. IV. as it evidently is to the eye, as any one may satisfy himself by inspecting the two figures.

The icosahedron just described, occurs among the secondary crystals of iron-pyrites. Naturalists at first were disposed to consider this as the regular geometrical icosahedron. But it has been demonstrated by HAUY, that the regular icosahedron does not exist among crystals, and cannot be produced by any law of decrement what ever. The same remark applies to the dodecahedron of mathematicians, a solid bounded by twelve regular and equal pentagons. No such crystal exists, nor can be produced by any law of decrement whatever. Of the five regular solids of mathematicians, the cube, the tetrahedron, the octahedron, the dodecahedron, and the ico-

sahedron,

sahedron, the first three occur in the mineral kingdom,
but not the last two.

It will be worth while to give another example of a
compound secondary form; and we shall take for that
purpose the regular six-sided prism of calcareous-spar,
Fig. 1. Pl. III. From the account formerly given of the
manner of dissecting this prism, it is easy to conceive that
its rhomboidal nucleus AA′, Fig. 5. has its solid lateral
angles E, O, I, K, G, H, situated in the middle of the
faces of the prisms; from which it follows, that these
angles are the points from which the decrements set out
that produce these faces.

These decrements act at once upon the three plane
angles EOI, EOA′, IOA′; but we may satisfy ourselves
with considering the decrement relative to one of these
angles, supposing the face which results from it extends
itself upon the two adjacent rhombs belonging to the
same angle. Let us agree, therefore, to restrict the
whole to the six angles EOI, EHG, IKG, HGK, OIK,
HGO, the three first of which are turned towards the
summit A, and the three last to the summit A. If
we suppose a decrement by two ranges of rhomboidal
molecules on these different angles, six faces will be
produced parallel to the axis, as has been already observed.

The plates of superposition, at the same time that they
undergo a decrement towards their inferior angles, will
extend by their superior parts so as to remain always
contiguous to the axis, the length of which will progressively augment. The faces produced by the decrement
will gradually increase, and when they touch each other
we shall have the solid AA′, Fig. 4. where each of the
faces, as o O o, is marked by the same letter as the angle

X   O

O, Fig. 5. to which it belongs, and which is now situated in the middle of the triangle o O o, because it constitutes the common point from which the three decrements set out.

In proportion as new plates are applied after this to the preceding ones, the points o, o rise up, while the point O sinks down, so that at a certain period we shall have the solid represented by Fig. 3. where the faces produced by the decrements are become pentagons, such as o o i O c.

Things being in this state, let us suppose a second decrement to concur with the first, and to take place by a single range upon the superior angle EAI, Fig. 5. and its opposite angle HA'K, always with this condition, that the face produced by it on both ends of the figure is continued upon the two rhombs adjacent to that to which the angles EAI, HA K belong. The effect of this decrement will be, to produce two faces perpendicular to the axis; and when it has reached the point at which these faces cut the six faces parallel to the axis produced by the first decrement, the secondary solid will be completed, and will be a regular six-sided prism Fig. 1. Pl. III.

We have already said that this result is general, whatever be the measure of the angles of the primitive rhomboid. We now see why, in the mechanical division of the prism, the cut p p o o, Fig. 2. Pl. III. has its sides p p, o o parallel to each other, and to the horizontal diagonal EI, Fig. 5. Pl. III.; since the two decrements taking place, the one upon the angle EOI, the other upon the angle EAI, the plates of superposition ought to have their edges turned towards this same diagonal.

In the case which we have been considering and which is the most usual, the axis of the secondary crystal is
longer

longer than that of the nucleus ; so that this nucleus having its lateral angles contiguous to the faces of the
prism, its summits are inclosed within the prism, at a
certain distance above the centre of the bases. If we
were to suppose that the two decrements began at the
same time, in that case the axis of the prism would be
equal to that of the nucleus, and the lateral angles and
summits of the nucleus would be tangents, the one to the
faces of the prism, the other to its bases. If the decrements on the superior angles of the nucleus were anterior
to the other decrements, which is the opposite of the first
case, the summits of the nucleus would then be contiguous to the bases of the prism, while its lateral angles
would be wholly within the prism, between the axis and
the prismatic faces. This is the case with certain crystals
in which the prism is very short, and resembles an hexagonal plate.

*Of Secondary Forms, when the Molecules differ from Paral-
lelopipeds.*

It is a character common to all the primitive forms,
to be divisible parallel to their different faces. In the
parallelopiped, when it is not joined by some other in
a different direction, such a division leads us obviously
to the form of a molecule, similar to that of the primitive
crystal. In the regular six-sided prism, it gives us for a
molecule a triangular equilateral prism. In the octahedron, it appears to produce molecules of two different
forms, some by tetrahedrons and octahedrons : the same
thing happens with respect to the tetrahedron. Various
ideas have been suggested by philosophers to get over the
difficulty in this case. Dr WOLLASTON has got rid of it
by supposing the molecules to be spherical, and to produce

duce the tetrahedrons and octahedrons, by combining of
fours and sixes. HAUY conceives that the tetrahedron is
the integrant molecule, and that the octahedrons are nothing else than empty spaces between the molecules, produced by these molecules uniting by their angles. The
subject does not admit of decision ; but as it is of no consequence to the theory of crystallography what opinion
we adopt, there is no occasion to enter upon the discus-
sion of the subject here. The rhomboidal dodecahedron, when divided in this manner, gives tetrahedrons
of isosceles triangular faces, equal and similar to each
other.

With respect to the dodecahedron with isosceles triangular faces, we cannot extract its integrant molecules
without dividing it in directions different from those
which are parallel to the faces. The cutting planes
must pass through the axis, and through the edges contiguous to angles of the summit. The molecules obtained are irregular tetrahedrons. The other primitive forms sometimes admit of division in directions parallel to the faces. This is the case with the rhomboid,
which constitutes the primitive form of the tourmaline.
It may be divided by planes passing through the axis
and the oblique diagonals. The result is the production
of tetrahedral molecules, such as are represented in
Fig. 33. Pl. IV.

Thus, besides parallelopipeds, there are two other
shapes which the integrant molecules assume ; namely,
the tetrahedron, and the triangular prism. Now, it de-
serves attention, and it is a point of considerable consequence in the theory of crystals, that the tetrahedral and
prismatic molecules are always arranged in such a manner in the interior of primitive and secondary crystals,
that,

that, taking them in groups of 2, 4, 6, or 8, they compose parallelopipeds; so that the ranges subtracted by
the effect of decrement, are nothing else than these parallelopipeds.

In order to conceive the better how this may be, let
us suppose for an instant that the molecules of calcareous-spar are divisible into tetrahedrons, as is the case
with the rhomboid, which constitutes the primitive form
of the tourmaline. This supposition will change nothing in the explanation of the different forms of which
calcarsous-spar is susceptible; that is to say, that in
determining the forms of this mineral, aided by the
theory, we may always satisfy ourselves with considering decrements by one or more ranges of rhomboidal molecules.

What is only a hypothesis with respect to calcareous
spar, is a reality with regard to the tourmaline. But
although the rhomboids, to which we arrive by mechanical division in this species, are themselves divisible
into tetrahedrons, still the decrements which produce
the secondary forms take place by the subtraction of
rhomboids similar to the primitive form; so that we
may suppose, in the calculations relative to the determination of these forms, that the tetrahedrons which constitute the true molecules are united together in an invari-
able manner, in each rhomboid.

Let us take another example from those crystals whose
primitive form is a regular six-sided prism. Let AD, Fig.
32. Pl. IV. be the base of such a prism, divided into small
triangles, which constitute the bases of the integrant molecules. It is evident, that any two neighbouring triangles whatever, such as A p i, AO i, compose a rhomb,
and of course the two prisms to which they belong form

by

by their union a prism with a rhomboidal base, which is
a species of parallelopiped. If we conceive that the two
triangular prisms, which constitute elements of the parallelopipeds, are invariably united together, it is obvious
that we may considered the six-sided prism as composed of
rhomboids instead of triangular prisms. Now, if we conceive a series of plates piled upon the hexagon ABCDFG
Fig. 32. which undergo, for example, upon their different edges, a subtraction of one row of parallelopipeds
similar to those that we are supposing here, these edges
will successively correspond with the lines of the hexagon
$i l m n r h$, $k u \kappa y g e$, &c., from which we see, that the
quantity by which each plate passes the other is a sum of
parallelopipeds or prisms with rhomboidal bases; and
it is easy to judge, that the result of the decrement,
supposing it to reach its limit, will be a right hexangu-
lar pyramid, which will have for its base the hexagon
ABCDFG.

All the other primitive forms different from the parallelopiped, give analogous results. We might even substitute for each of these forms a nucleus similar to the
little parallelopipeds, which are formed by the union of
the tetrahedrons or triangular prisms, and we would succeed equally in explaining the secondary forms by laws
of decrement applied to that nucleus, which would be obtained likewise by mechanical division.

HAUY calls these parallelopipeds composed of tetrahedrons or triangular prisms, *subtractive molecules*. They
are always substituted in place of the, tetrahedrons or
triangular prisms, in considering the decrements which
produce the secondary forms in these cases. Thus, as
far as the theory of crystals is concerned, we have
                                                      nothing

nothing to do with the integrant molecules, but may conceive all crystals composed of a congeries of parallelopipeds.

4. *Difference between Structure and Increment.*

In the preceding exposition of the Theory of Crystallography, it has been supposed that the laminæ of which the crystals of the same species are composed, proceed from a commom nucleus, undergoing decrements subjected to certain laws, upon which the forms of these crystals depend. But this is only a conception adopted in order to make us more easily perceive the mutual relations of the form in question. Properly speaking, a crystal is only a regular group of similar molecules. It does not commence by a nucleus of a size proportioned to the volume which it ought to acquire, or, what comes to the same thing, by a nucleus equal to that which we extract by the aid of mechanical division ; and the laminæ which cover this nucleus, are not applied successively upon each other in the same order in which the theory regards them.

The proof of this is, that among crystals of different sizes, which are frequently attached to the same basis, the most minute are as perfectly formed as the largest ; from which it follows, that they have the same structure, that is to say, they already contain a small nucleus, proportioned to their diameter, and enveloped by a number of decreasing laminæ, necessary in order that the polyhedron should be provided with all its faces. We do not perceive these various transitions from the primitive to the secondary forms, which, however, ought to be discoverable, if, during the process of crystallisation, the pyramids

ramids resting on the nucleus were formed progressively
in layers from the base to the apex. This, however, is
only true in general ; for it sometimes happens, in artificial crystallisation, (and it is probable that it also occurs
in natural bodies,) that a form, which had attained to a
certain size, suddenly experiences variations by the effect
of some particular circumstance. We must therefore
conceive, for example, that from the first moment a crystal, similar to the rhomboidal dodecahedron, is already a
very small dodecahedron, which contains a cubical nucleus, proportionally small, and that this kind of embryo
continues to increase, without changing its form, by the
addition of new laminæ on all sides ; so that the nucleus
increases on its part, always preserving the same relation
with the entire crystal.

We shall make this idea more distinct by a construction, which refers to the dodecahedron already mentioned, and represented by means of a plane figure. What
we shall say of this figure, may be easily applied to a solid, since we may always conceive a plane figure, like a
section made in a solid.

Let $t s z s'$, Fig. 54. A, Pl. V. be an assortment of
small squares, in which the square BNDG, composed of
49 imperfect squares, represents the section of the nucleus, and the extreme squares t, p, i, B, f, c, s, &c-
that of the kind of steps formed by the laminæ of superposition. We may conceive, that the assortment has
commenced by the square BNDG, and that different
piles of small squares are afterwards applied on each
of the sides of the central square ; for example, on the
side BN, at first the five squares comprehended between f and h, afterwards the three squares, contained
between

between *c* and *e*, and then the square *e*. This progress corresponds with what would take place if the dodecahedron commenced with a cube proportioned to its volume, and which afterwards increased by an addition of laminæ continually decreasing

But, on the other hand, we may imagine that the assortment had been at first similar to that which is represented by Fig. 54. C. Pl. V., in which the square BNDG is only composed of nine molecules, and had on each of its sides only a single square *s, t, s', z*. If we refer, in imagination, this assortment to the solid of which A is the section, we shall easily judge that this solid had for its nucleus a cube composed of 27 molecules, and of which each face, composed of nine squares, carried on that of the middle a small cube, so that the decrement of one range is already exhibited in this initial dodecahedron. By application of new squares, this assortment will become that of Fig 54. B. Pl. V. in which the central square BNDG is formed of 25 small squares, and supports on each of its sides a pile of three squares, besides a terminal square *s, t, s'*, or *z*. Here we have already two laminæ of superposition instead of one only. Finally, by an ulterior application, the assortment of Fig. 54. B. Pl. V. will be changed into that of Fig. 54. A. Pl. V. where we see three laminæ of superposition.

These different transitions, of which we are at liberty to continue the series as far as we please, will give an idea of the manner in which secondary crystals may increase in magnitude, and still preserve their form ; from which we may judge that the structure keeps pace with this augmentation of volume ; so that the law, according to which all the laminæ are applied on the nucleus, when it

Y   has

has attained its greatest diameter, and which successively decrease, was already sketched in the nascent crystal.

5. *Of those Crystals in which one half is turned round, and
of others that intersect each other.*

Having considered the most perfect and regular form
of crystals, we shall now speak of certain accidents, which,
under the appearance of exceptions or anomalies, still
possess a latent tendency toward the same laws to which
the structure is subjected, when nothing deranges their
progress, or disturbs their harmony.

In ordinary crystals, the faces adjacent to each other
always form salient, and never re-entering angles.   But
crystalline forms also exist which present these last angles ; and ROME' DE LISLE was the first who observed,
that this effect took place when one of the two halves of
a crystal was in a reversed position with respect to the
other.   A very simple example will enable us to conceive
this reversed position.

Let us suppose that B d, Fig 55. Pl. VI. represents an
oblique prism, with rhombic bases, situated in such a
manner, that the faces AD d a, CD d c, are vertical, and
BD are the acute angles of the base ; and the latter proceeds in an ascending direction from A towards C.   Let
us also suppose, that the prism is cut into two halves, by
means of a plane which should pass by the diagonals
drawn from B to D, and from b to d, and that the half
situated on the left, remaining fixed, the other is reversed, without being separated from the former.   The
crystal will be presented under the aspect which we see
in Fig. 56. Pl. VI., where the triangle b' d' c', which was
one

one of the halves of the inferior base, Fig. 55. Pl. VI.
is now situated on the upper part, Fig. 56. Pl. VI. and
forms a salient angle with the fixed triangle ABD ; while
the triangle BDC, Fig. 56. Pl. VI. which was one of the
halves of the superior base, Fig. 55. Pl. VI. is transported into the lower part, Fig. 56. Pl. VI. and forms a
re-entering angle with the fixed triangle a b d.

We can easily conceive that the plane of junction
DB b d of the two halves of a rhomboid, is situated like
a plane produced by a decrement of one range on one or
other of the edges A a, C c, Fig. 55. Pl. VI. ; and thus
the manner in which these halves join, is in strict relation
with the structure.

Now if we imagine a secondary form, which has for its
nucleus the same prism, and if we suppose that it has
been cut in the direction of the plane DB b d, and that
one of its halves is reversed in such a manner, that the
half of the nucleus which corresponds with it, assumes
the same position as in the preceding case, the arrangement might be such, that there is still a re-entering angle
on the one hand, and a salient angle on the other, which
will result from the mutual incidences of the faces produced by the decrements.

In certain cases, the plane of junction, on which the
two halves of the crystal are joined, is situated parallel
to one of the faces of the nucleus and the arrangement
does not admit of presenting a re-entering angle opposite
to a salient angle.

Haüy has given to these reversed crystals the name
hemi-tropes, denoting one half reversed.

Another accident, extremely common, is the manner
in which grouped crystals are inserted into each other.
This kind of apparent penetration is subject to so many
diversities,

diversities, that frequently, among crystals of the same group, we do not find two relative positions resembling each other. But although, in general, the positions in grouped crystals are extremely variable, we find, on more accurate examination, that they are subject to certain laws always analogous to those of structure.

Let us take a simple example to illustrate this. Let AC, Fig. 57. Pl. VI. be a cube, and MN r an equilateral triangular facet, produced in place of the angle A, in consequence of a decrement by one range round this same angle. Let us suppose a second cube modified in the same manner, and attached to the former by the facet which results from the decrement indicated by M, N, r. We shall have the assortment represented by Fig. 58. Pl. VI.

We may now conceive that one of the two cubes, that, for example, which is placed below, is increased in all its dimensions, except at the places where the other forms an obstacle to it. In proportion as this increment becomes more considerable, the upper cube will be more and more enveloped in the inferior one, and it may even finish by being entirely concealed by it. We observe crystals sunk into each other all degrees of depth, but always in such a manner, that their plane of junction has a position analogous to planes resulting from decrement, so that both structures follow their ordinary progress, to this plane, which serves as their mutual limit or boundary. Haüy having divided cubes of fluor-spar inserted into each other, remarked, that the folia of each, extended without interruption, until stopped by the common plane of junction.

The example just mentioned relates to a very simple and very regular law of decrement. But frequently the
laws

laws which determine the plane of junction are more or
less complicated, and there are a few which are somewhat
extraordinary.

When two prisms cross towards the middle of their
axis, there are two planes of junction which unite, crossing each other, as in the mineral named *grenatite*, Fig. 63.
Pl. III. of *System of Mineralogy*, and these planes have
positions analogous to those which would be determined
immediately by the known laws of decrement.

6. *Of the Symbols used to denote the particular Laws of Decrement which produce the Secondary Forms.*

HAUY has invented particular symbols, to denote the
particular laws of decrement which produce the secondary forms. As these symbols occur constantly not
only in his writings, but also in those of all the authors
of the same school of mineralogy, and as they are useful
by greatly shortening the account of the formation of secondary crystals, we shall here explain them.

Let Fig. 59. Pl. VI. represent an oblique parallelopiped,
the faces of which have angles with different measures,
and let it be the primitive form of some mineral ; as, for
example, of felspar.

The vowels are adopted to represent the solid angles.
The four first, A, E, I, O, are placed at the four angles
of the superior base following the order of the alphabet,
and that of ordinary writing, namely, beginning at the
top, and going from left to right.

The consonants are chosen to denote the edges   The
six first, B, C, D, F, G, H, are placed on the middle of
the edges of the superior base, and upon the two longitudinal edges of the lateral faces, which occur first in going
from

from left to right. These consonants are likewise arranged in the alphabetical order, ana according to the usual
mode of writing.

The letters P, M, T, which are the initials of the syl-
lables of which the word *primitive* is composed, are placed each in the middle of the superior base, and of the two
lateral faces exhibited to view.

Each of the four solid angles, or of the six edges marked by letters, is susceptible in the present case, on account
of the irregular form of the parallelopiped, of undergoing
particular laws of decrement. Hence the reason why
they are marked each with a different letter. But as the
laws of decrement act with the greatest symmetry possible, every thing which takes place with respect to the
angles and edges marked with letters, takes place also
with respect to the opposite angles and edges which are
not marked, or are not visible. It was only necessary to
mark the number of solid angles or edges which undergo
distinct decrements, because these decrements include
likewise implicitly all those which take place upon analogous angles or edges.

In some cases, however, it is necessary to indicate these
last angles or edges. In such cases, the small letters,
having the same names as the capitals, are employed for
the purpose. The angles analogous to A, E, I, O, are
denoted by a, e, i, o; and the edges analogous to B, C,
D, F, G, H, are denoted by b, c, d, f, g, h. But it is
very seldom necessary to mark these small letters on the
figure ; it is sufficient to introduce them into the symbol
of the crystal, because it is easy to conceive the place
which every one ought to occupy in the figure.

To indicate the effects of decrements by one, two,
three, four, or more ranges in breadth, the figures 1, 2,
3,

3, 4, &c. are employed in the way to be immediately explained; and, to indicate the effects of decrements by 2,
3, &c. ranges in height, the fractions $\frac{1}{2}$, $\frac{1}{3}$, $\frac{1}{4}$, &c. are employed.

The three letters P, M, T, serve to distinguish either
the form of the nucleus, without any modification, when
they alone constitute the symbol of the crystal, or the faces parallel to those of the nucleus, in the case where the
decrements do not reach their limit; and then these letters are combined in the symbol of the crystal with those
which relate to the angles or edges that have undergone
decrements.

Let us suppose, at first, for the greater simplicity, that
one of the solid angles, such as O, is intercepted by a
single additional face. The decrement which produces
this face may take place either on the base P., or on the
face T, which is on the right of the observer; or on the
face M, which is on the left. In the first case, the figure marking the decrement is placed above the letter O;
in the second case, the figure is placed like an ordinary
exponent; in the third case, it is placed on the left side,
and somewhat above the letter.

Thus, $\overset{2}{O}$ denotes the effect of a decrement by two ranges in breadth, parallel to the diagonal of the base P,
which passes through the angle E. O^3 indicates the effect of a decrement by three ranges in breadth, parallel
to the diagonal of the face T, which passes through the
angle I. ^4O indicates the effect of a decrement by four
ranges in breadth, parallel to the diagonal of the face M
that passes through the angle E.

When the decrement relates to some one of the three
other solid angles I, A, E, the observer is conceived to
move

move round the crystal till he is opposite to that angle, as he is naturally opposite to the angle O in the case which we have been describing ; or, what comes to the same thing, he is conceived to turn round the crystal till the solid angle that he is considering be exactly opposite to him, and it is relative to that position that a decrement is said to take place towards the right or towards the left.

For example, if we are speaking of the solid angle A, the sign A^2 will represent the effect of a decrement by two ranges on the surface AE s r, Fig. 60. Pl. VI. or upon that which is opposite to T, Fig. 59. Pl VI. ; and ^5A will represent the effect of a decrement by three ranges upon the face AI u r, Fig. 60. Pl. VI. or upon that which is opposite to M, Fig. 59. Pl. VI.

As to the decrements on the edges, those which take place towards the boundary BCFD of the base, are expressed by a number placed above or below the letter, according as their effect takes place in going up or going down, supposing them to set out from the edge to which they are referred ; while those which take place on the longitudinal edges G, H, are indicated by an exponent placed on the right or the left of the letter, according as they take place in one direction or the other. Thus $\overset{2}{D}$ expresses a decrement by two ranges proceeding from D towards C ; $\overset{3}{C}$ a decrement by three ranges going from C towards D ; D a decrement by two ranges, descending upon the face M $\overset{2}{}$ ^5H a decrement by three ranges, proceeding from H towards G ; ^4G a decrement by four ranges, proceeding from G towards the edge opposite to H, &c.

When

When it is necessary to denote by a small letter, such
as d, a decrement upon the edge $u\,r$, Fig. 60. Pl. VI. opposite to the edge denoted by the capital letter D, Fig.
59. Pl. VI., we must suppose the crystal turned upside
down.   Hence $\overset{\scriptscriptstyle 2}{d}$ will express a decrement by two ranges
upon the base p, just as $\overset{\scriptscriptstyle 2}{D}$ expresses a similar decrement
on the base P.   For the same reason, $\underset{\scriptscriptstyle 3}{c}$ will express a
decrement by three ranges, proceeding from $s\,p$ towards
EO, Fig. 60. Pl. VI.

If the same solid angle, or the same edge, undergo several successive decrements on the same side, or different
decrements which take place on different sides, in that
case, the letter pointing out the angle or edge is repeated
as often as the decrements, varying the figure each time,
to make it correspond with the particular decrements
pointed out.   Thus, $\overset{\scriptscriptstyle 2}{D}\underset{\scriptscriptstyle 3}{D}$ will denote two decrements up-
upon the edge D, one of two ranges upon the base P.
another of three ranges upon the face M. $^2H\,^4H$ will
denote two decrements, the one by two ranges, the other
by four, on the left of the edge H.

Mixed decrements are marked according to the same
principles, employing the fractions $\frac{2}{3}$, $\frac{3}{4}$, &c. which represent them ; the numerator referring to the decrements
in breadth, and the denominator to decrements in
height.

The method of describing the intermediate decrements
still remains to be explained.   This will be best done by
an example.   Let AEOI, Fig. 61. Pl. VI. be the same
face as in Fig. 59. Pl. VI.   Let us suppose a decrement
by one range of double molecules, according to lines parallel to $x\,y$, so that Oy measures the double length of a
molecule,

molecule, O x lines equal to a single molecule. This kind of decrement is written in this manner, ($\overset{1}{O}$ D¹ F²). The parenthesis lets us know, in the first place, that the decrement is intermediate ; $\overset{1}{O}$ indicates that it takes place by one range upon the angle marked by that letter ; and that it belongs to the base AEOI, Fig. 59. Pl. VI. D¹ F² indicates that there is one length of a molecule taken away along the edge D, and two lengths along the edge F.

It is useful to have a language to denote these symbols, so that they may be easily written down when dictated by another person. On that account, we shall mention here the mode followed for that purpose. The symbols O², ³O, are thus read: O *two on the right*, O *three on the left*, $\overset{2}{O}$, O thus, O *under two*, O *above four*.

Finally, the symbol ($\overset{1}{O}$ D¹ F²) thus, *in a parenthesis*, O *under one*, D *one*, F *two*.

We must now notice the order in which these letters must be placed, in order to denote a secondary crystal. If the alphabetical order were adopted, there would result a sort of confusion in the picture which the formula presents. It is more natural to conform to the order which would direct an observer in the description of the crystal; that is to say, to begin with the prism or the middle part, and to indicate its different faces as they present themselves successively to the eye ; then to pass to the faces of the summit or the pyramid.

Suppose, now, that Fig. 62. Pl. VI. represents the *bilinary* variety of felspar, the primitive form of which is seen in Fig. 59. Pl. VI. In this va-

riety,

riety, the face *l*, Fig. 62. Pl. VI. results from a decrement by two ranges on the edge G, Fig. 59. Pl. VI. going towards H. The face M, Fig. 62. Pl. VI. corresponds with that which is marked with the same letter in
Fig. 59. Pl. VI. and which is only concealed in part by
the effect of the decrement. The face T, Fig. 62. Pl.
VI. is parallel to T, Fig 59. Pl. VI. The pentagon *x*,
Fig. 62. Pl. VI. comes from a decrement by two ranges
on the angle I, Fig. 59. Pl. VI. parallel to the diagonal
AO. As this decrement does not reach its limit, the
summit exhibits a second pentagon P, Fig. 62. Pl. VI.
parallel to the base P, Fig. 59. Pl. VI. All this description may be exhibited in symbolic language, as follows :

$$G^2\ M\ T\ \overset{2}{I}\ P.$$

In order to prevent beginners from finding any thing
ambiguous in this symbolical mode of writing, especi-
ally in complicated cases, Haüy is in the habit of placing under the different letters which compose the symbol, those which correspond to them in the figure. If
we adopt this mode, which is a considerable improve-
ment, the symbol denoting *bibinary* felspar will be as

follows: $G^2 M\ T\ \overset{2}{I}\ P.$
$l\ M\ T\ x\ P.$

These letters thus written below, enable us to compare
the symbol with the figure, and thus to decypher the
meaning with facility, how complicated soever it should
be But some more observations will be necessary, in order to understand fully the way in which these symbols
are employed,

Let us now, then, turn our attention to parallelopipeds
of a more regular form than that which constitutes the

primitive

primitive crystal of felspar.   But let us suppose them at
first not to be rhomboids.   They are nothing else than
what is represented in Fig. 59. Pl. VI. but the form has
varied, so as to render them symmetrical   In consequence of this alteration, certain angles and edges,
which differed from each other in the first parallelopiped,
have become equal in this.   Hence, every thing that takes
place on one of them is repeated on the other.   They
ought, therefore, to be denoted by the same letter.   Thus,
in algebra, certain general solutions are simplified in particular cases, when a quantity, at first supposed to be different from another, becomes equal to it.

Let us suppose, for example, that the primitive form
is a rectangular prism, having oblique angled parallelo-
grams for its bases, one side of which is longer than
the other.   In that case, we have $O=A$, Fig. 59 Pl. VI.
$I=E$, &c.   In such a case, the first letter of the alphabet will be substituted for the other, as is done in Fig.
63. Pl. VI.

If we pass through the different kinds of parallelopipeds, we shall find them acquire different degrees of simplicity, which occasions new equalities in the angles and
edges, and of course new substitutions of letters.   We
shall have successively,

The oblique prism with rhomboidal bases, respresented
in Fig. 19. Pl. III.

The rectangular prism, with rectangular bases, represented in Fig. 16. Pl. III.

The rectangular prism, with rhomboidal bases, represented in Fig. 17. Pl. III.

The rectangular prism, with square bases, represented
in Fig. 15. Pl. III.

The

The cube represented in Fig. 63. Pl.VI. Here only the superior base is marked with letters, because what takes place with respect to it may be applied indifferently to any of the other faces.

The same mode is followed in writing the symbols for these different forms, only the letters that have the same name and the same figures, are not repeated. An example will render the method evident. Fig. 64. Pl VI. represents the most common variety of the *crysoberyl*, the nucleus of which is a rectangular parallelopiped, such as is represented in Fig. 65. Pl. VI. The symbol

of the secondary crystal will be $M T \; {}^{2}G \; G^{2} \; \overset{1}{B} A^{\frac{3}{2} \frac{3}{2}}A.$
$\qquad\qquad\qquad M T \quad s \quad i \quad o$

This variety is called Annular *Crysoberyl*

To understand the preceding expression better, let us mark each angle and edge with a particular letter, as if the parallelopiped were oblique angled. See Fig. 65 Pl. VI.

In that case, the symbol would become $MT \; {}^{2}G \; H^{2 \cdot} \overset{1}{B} \overset{1}{F}$
$E^{\frac{3}{2} \frac{3}{2}}O.$ But if we compare Fig. 65 ª with Fig. 65. we see, that H =G, F B, O=A, &c. Hence, if we substitute, instead of the first letters, their values, we get
$MT \; {}^{2}G \; G^{2} \; \overset{1}{B} \overset{1}{B} A^{\frac{3}{2} \frac{3}{2}}A,$ which becomes the same with the one given above, when the useless repetition of $\overset{1}{B}$ is suppressed.

From the preceding statement, it is evident that we must take care not to confound, for example, ${}^{2}G \; G^{2}$ with $G^{2} \; {}^{2}G.$ The first symbol indicates the decrements which take place on the face T, Fig. 65. Pl VI. and on the face opposite to it, going from the edges G towards those that correspond with them behind the parallelopiped The second indicates the decrements which take place upon the face

M,

M, and which meet each other in the middle of that face.
If these two decrements took place simultaneously, their
symbol would be $^2G^2$.

In the preceding symbols, each letter, such as ^2G or
G^2 can only be applied to a single edge, situated to the
right or the left, as that letter itself. But $^2G^2$ applies
indifferently to the one edge or the other. Hence, it is
needless to repeat that letter.

Let us take the Figure 66. as another example *. If
we suppose Fig. 17. to represent its primitive form, we
will have for the symbol of the variety of crystal here represented, $^3G^3$ M $\overset{2}{B}\overset{3}{B}\overset{1}{E}\overset{2}{E}$ P.

$\quad\quad\quad$ o M r s z u P.

In this symbol $^3G^3$ indicates two distinct faces formed on each side of each edge G. But it is not necessary
to place two letters under that symbol, because all the
faces situated in the same manner being distinguished by
the same letter in the figure, it is sufficient to point out
that the symbol $^3G^3$ applies to the faces marked with
the letter o, and this requires only to write the letter o,
once under the symbol.

From the same principles, it follows, that the rhomboidal dodecahedron derived from the cube, Fig. 63. Pl.
VI. is expressed by the symbol $\overset{1}{B}$ B. The octahedron
derived from the cube is expressed thus, $\overset{1}{A}$ $^1A^1$.

The rhomboid, supposing it placed in the most natural
aspect ; that is to say, so that the two solid angles com-
$\quad\quad\quad\quad\quad\quad\quad\quad\quad\quad\quad\quad\quad\quad$ posed

---

* This figure represents a variety of the topaz ; of course, our supposi-
tiun respecting the primitive crystal is not accurate. But that does not in-
jure the illustration.

posed of three equal plane angles, are in the same vertical line, has, properly speaking, no base, but merely sum.
mits, which are the extremities of its axis. Its angles
and edges are marked as in Fig. 67. Pl. VI. The letter
e denotes that the angle marked by it is similar to that
which is marked with a capital E. So that if all the lateral angles were indicated by letters, the three nearest
the superior summit, would have the letter E, and
the three nearest the inferior summit the letter *e*.

As the rhomboid has its six faces equal and similar, it
is only necessary to consider the decrements relative to
one of these faces ; as, for example, the one which in the
figure is marked P, because all the others are mere repe-
titions of this. These observations suggest the following
rules : 1. The decrements which set out from the superior angle A, or the superior edge B, will have the figure
indicating the number of ranges placed below A and B.
2. Those which set out from the lateral angles E will
have their figures situated at the side and towards the
top of the same letter. 3. With respect to those which
set out from the inferior angle *e*, or from the inferior edge
D, the figure will be placed above the letter *e* or D.

Suppose, for example, that Fig. 68. Pl. VI. represents
the variety of calcareous-spar called *analogic* by HAUY, its
symbol will be $\overset{v\ g}{e}$ D B.
$$\underset{c\ r\ g}{\overset{\ \ 1}{}}$$

What has been said of the rhomboid is easily applied
to the other primitive forms. Put probably some illus
trations will be considered as necessary to make the symbols applied to them the more readily understood. On
that account we shall take a short review of each of
them.

Fig.

Fig.69. Pl. VI. represents the octahedron with scalene triangles, Fig. 71. Pl. V1. the octahedron with isosceles triangles, and Fig. 70. Pl. VI. the regular octahedron.

In placing the figures which accompany the letters in the symbols, the same rule is followed that was described with respect to the rhomboid. Thus, in Fig. 70. Pl. VI. the figure is placed below the letter to represent the decrements setting out from the angle A or the edge B; it is placed above for those which set out from the edge D; and at the side, for those which set out from the angle E. If we want to denote the result of a decrement by one range upon all the angles of the regular octahedron, Fig. 71. Pl. VI. we have only to write $\overset{1}{A}$ $^1A^1$. To indicate the result of a decrement by one range on all the edges, we write B $\overset{1}{B}$. The first of these decrements produces a cube, the second a rhomboidal dodecahedron.

In some species, as in the nitrate of potash, the primitive octahedron, the surface of which is composed of eight isosceles triangles, similar 4 and 4 to each other, ought to have the position represented in Fig. 72. Pl. VI. that the secondary crystals may have the most natural attitude; that is to say, that the edges which join the two pyramids which compose the octahedron, ought to be two of them in a vertical direction, as F, and two in a horizontal, as B. By comparing Fig. 71. Pl. VI. with Fig. 72 Pl. VI. in which the letters are placed as if all the angles and edges had different functions, it will be easy to conceive the distribution adopted in Fig. 71. Pl. VI. and brought to the symmetry of the true primitive form. For, in the present case, we have E=A, D=C, G=F.

The

The figure denoting the number of ranges, will be placed under the letter, to denote decrements proceeding from B. It will be placed at one side, or below, to denote those proceeding from A; according as their effect respects the triangle AIA, or the triangle AIF. It will be placed above or below, for those which proceed from C, according as their effect is produced on the first or the second of these triangles. It will be placed at one side for the decrements which proceed from F. Finally, it will be placed above, below, or on either side, for the decrements that proceed from I, according as their effect takes place towards B or towards F.

The tetrahedron being always regular, when it becomes the primitive form, it will be expressed as in Fig. 74. Pl. VI. To indicate, for example, a decrement by three ranges on all the edges, we would write B $\overset{3}{\text{B}}$; and to indicate a decrement by two ranges upon all the angles, we would write A $\underset{2}{^2\text{A}^2}$, as in the case of the regular octahedron.

A simple inspection of Fig. 75. Pl. VI. is sufficient to make us understand the symbols in the case of regular six-sided prisms. The figures are written precisely in the manner already described for the four sided prism; to which, therefore, we refer the reader. But it happens sometimes that three of the solid angles taken alternately are replaced by faces, while the intermediate angles remain untouched. In that case the prism is distinguished as in Fig. 76. Pl. VII.

In the rhomboidal dodecahedron, Fig. 77. Pl. VII. each solid angle composed of three planes may be assimilated to a summit of the obtuse rhomboid. Hence, it is only

A a   necessary

necessary to give letters to one face, as may be seen in the figure.

Hitherto there has been no occasion to use any symbols for the dodecahedron with triangular faces, because it is more natural to substitute in place of it the rhomboid from which it is derived, and which gives simpler laws of decrement.

We have still to explain the method of representing a peculiar case, which sometimes occurs in some crystals, where the parts opposite to those which undergo certain decrements remain untouched, or are modified by different laws. This case belongs chiefly to the tourmaline, and it is easy to indicate its peculiarity by means of zeros.

For example, in the variety of tourmaline represented in Fig. 79. Pl. VII. the primitive form of which is represented in Fig. 7S. Pl. VII.; the prism, which is nine sided, has six of its faces, namely, s, s, Fig. 79. Pl. VII. produced by the subtraction of one range upon the edges D, D, Fig. 78. Pl. VII. and the three others, such as l, by the subtraction of two ranges on the three angles e, Fig. 78. Pl. VII. only. Farther, the inferior summit has only three faces parallel to those of the nucleus; while, on the superior summit the three edges B are replaced each by a face n, n, Fig 79. Pl. VII. in consequence of a decrement which has not reached its limit. This crystal is represented by the following symbol: $\overset{1}{D}\overset{2}{e}\overset{2 \cdot 0}{E}\ P\ B\ \underset{1}{b}$. The
$$\underset{1}{s}\ \underset{1 \cdot 0}{l}\quad P\ n$$

quantities $\overset{2\ 0}{E}$, $\underset{1\ 0}{b}$ indicate, the one that the angles E, Fig.78. Pl. VII. opposite to e undergo no decrement; the other, that the edges parallel to B remain equally untouched.

If

If these edges underwent a different law, which produced, for example, an abstraction of two ranges, the
symbol would become $\overset{1}{D}\overset{2}{e}\overset{2}{E}\overset{0}{P}B\underset{1}{b}\underset{2}{}$. From this, it is
obvious, that it must be understood that the decrements
represented by a capital letter accompanied by any figure, do not implicitly include the similar decrements represented by a small letter of the same name, or the opposite, that is to say, that B does not implicitly include
$\underset{2}{b}$, or *vice versa*, except when the second letter does not
enter into the symbol with a different figure, or does not
bear the same figure accompanied by a zero. In the first
case, each of the two letters indicates a decrement which
is peculiar to the edge or angle indicated by it. In the
second case, the zero indicates that the angle or edge to
which it exclusively relates undergoes no decrement what-
ever Thus, in the symbol $\overset{1}{D}\overset{2}{e}\overset{2}{E}\overset{0}{P}B\underset{1}{b}\underset{2}{}$, B expresses a decrement by one range, which takes place only on
the edges contiguous to the superior summit A, Fig. 78.
Pl. VII. ; b indicates a decrement by two ranges, which
$\underset{2}{}$
only takes place on the edges contiguous to the inferior
summit. The quantities $\underset{2}{e}$ and $\overset{2..0}{E}$ ought likewise to be
considered independent of each other ; the first indica-
ting a decrement of two ranges on the angles e only,
and the second indicating that no decrement whatever
takes place upon the angles E, opposite to the preceding.

The foregoing observations have been given in considerable detail, in order to put our readers completely in
possession of the method, and to enable them to make a
figure

figure of a secondary crystal, merely from the symbol re,
presenting the laws of its formation. But to enable any
person to read these symbols, and to understand them,
much shorter directions would have sufficed. We shall
subjoin the following rules, which will be sufficient for
that purpose, and which will serve as a kind of epitome
of the preceding observations:

1. Every vowel employed in the symbol of a crystal
indicates a solid angle, marked with the same letter in
the figure which represents the nucleus. Every consonant indicates the edge which has the same letter in the
figure.

2. Each vowel and consonant is accompanied by a figure, the value and position of which indicate the law of
decrement which the corresponding angle or edge undergoes. We must except the three consonants P, M, T;
each of which, when it appears in the symbol of a crystal, indicates that the crystal has faces parallel to those
faces which have the same letters on the figure of the nucleus.

3. Each letter contained in the symbol of a crystal, is
understood, with the figure belonging to it, to apply to
all the angles or edges which have the same function as it
in the figure, and is marked with the same letter.

4. Every number joined to a letter indicates a decrement, setting out from the angle or the edge denoted by
that letter. If the number is a whole number, it indicates how many ranges in breadth are subtracted, supposing each plate to have only the thickness of one molecule. If the number is a fraction, the numerator indicates the number of ranges subtracted in breadth, and
the denominator the number of ranges subtracted in
height

5.

5. According as the number is placed below or above the letter which it accompanies, it indicates that the decrement descends or ascends, setting out from the angle or edge marked by the letter. If it is placed towards the top, and either on the right or the left side of the letter, it indicates a decrement in a lateral direction, either to the right or to the left of the angle or edge marked by the letter.

6. When a letter is twice repeated, with the same number placed on two different sides, as ^2G G^2 or G^2 ^2G, ^2A A^2 or A^2 ^2A, the two edges, or the two angles which it marks, ought to be considered on the figure in the same relative positions ; that is to say, for example, that in the symbol ^2G G^2, the quantity ^2G indicates the effect of decrement on the edge G situated at the left, and the quantity G^2 the effect of decrement upon the edge situated at the right.

7. When a letter has the same number both on the left and the right side, as $^3G^3$, it applies equally to all the edges G. The same thing holds with the letters which belong to the angles.

8. The parenthesis, as for example $(\overset{3}{O}D^1 \ F^2)$, indicates an intermediate decrement. The letter $\overset{3}{O}$ indicates, in the first place, that the decrement takes place by three ranges an the angle O, and that its effect is ascending. D^1 F^2 indicate, that for one molecule subtracted along the edge D, there are two molecules subtracted along the edge F.

9. Every small letter occurring in the symbol of a crystal, indicates the angle or the edge diametrically opposite to that which has the capital letter of the same name in the figure, where the small letter is omitted as superfluous.

superfluous. We must except the letter *e*, which is always employed in the rhomboid, and which indicates, according to the principle, the angle opposite to that which bears the letter E.

10. When a symbol contains two letters of the same name, the one large the other small, with different numbers attached to them, the two opposite edges or angles to which these letters belong, are conceived to undergo each exclusively the law of decrement indicated by the number attached to the letter.

11. Every letter, whether large or small, marked by a number having a zero following it, indicates that the decrement denoted by that number does not take place on the particular edge or angle denoted by the letter.

7. *The Nomenclature of Crystals.*

In the Wernerian Crystallography, the different regular forms in general are expressed by short descriptions. Haüy, on the contrary, has attempted to designate them by names taken from the characters of the forms themselves, or from the properties that result from their structure, and from the laws of decrement on which they depend. In this manner he constructed the following nomenclature.

I. *Primitive Forms.*

When a crystallisation has the same figure as the primitive nucleus, and is therefore a primitive form, Haüy designates it, by prefixing to the name of the mineral the word *primitive;* for example, *Primitive Zircon, Primitive Calcareous-spar, Primitive Iron-pyrites.*

II.

II. *Secondary Forms.*

These may be considered under the following points of view.

1. In relation to the modifications which the primitive form exhibits when its planes are associated with those originating from the laws of decrement.

2. Secondary forms, considered as purely geometrical forms.

3. In relation to certain planes or edges which are remarkable on account of their arrangement or position,

4. In relation to the laws of decrement on which they depend.

5. In relation to their geometrical properties.

6. Lastly, In relation to certain particular accidental circumstances.

1. *Secondary forms, considered in relation to the modifications which the primitive form exhibits, when its planes are associated with those resulting from the laws of decrement.*

A crystal is named

a. *Pyramidated (pyramidé),* when the primitive form is a prism, and has a pyramid on each extremity, with as many planes as the prism has lateral planes. Example, Pyramidated apatite, (chaux phosphatée pyramidie,) Fig. 130. *System of Mineralogy* *.

and

---

b. *Prismated* (prismé) when the primitive form is
   composed of two pyramids joined base to base,
   and the pyramids separated by a prism. Examples, Prismated zircon (zircon prismé) Fig. 3.
   Syst. Min. and prismated quartz (quartz prismé)
   Fig. 66. *Syst. Min.*

c. *Semi-prismated* (semi-prismé) when only the half
   of the edges on the common basis are obliterated
   by lateral planes. Example, Semi-prismated sulphat of lead (plomb sulphaté semi-prismé,)
   Fig. 241. *Syst. Min.* It is an elongated double
   four-sided pyramid, in which the two opposite
   edges of the common basis are truncated

d. *Based* (basé,) when the primitive form is either a
   double pyramid, or a rhomboid, in which the
   summits are intercepted by planes perpendicular
   to the axis, and which take the place of terminal planes. Example, Based sulphur (soufre
   basé,) Fig. 150. *Syst. Min.* It is a double
   four sided pyramid, truncated on the extremities.

e. *Epointé* , when all the solid angles of the primitive form are obliterated by single planes.
   Example, Radiated zeolite mesotype epointé.
   It is a four-sided prism, deeply truncated on all
   the angles; or, according to WERNER, a four-
   sided prism, acuminated on the extremities with
   four planes, the acuminating planes set on the
   lateral edges, and both the acuminations again
   truncated. The terms *bis-epointé, tri-epointé,
   quadri-epointé,* are used to express the angles
   being

---

* I here use the French word *epointé*, not recollecting any appropriate
English term.

OF MINERALS. 193

being intercepted by two, three or four planes.
Examples:

Analcime tri-epointé, Fig. 79. Pl. VI. which is
a cube acuminated on all the angles with
three planes.

Fer sulphuré quadri epointé: which is a cube
acuminated on all the angles, with three
planes and the summits of the acuminations
truncated.

f. *Emarginated*, (emarginée), when every edge of the
primitive form, is intercepted by a plane or facet.
Example, Emarginated garnet (grenat emarginé,)
Fig. 58. *Syst. Min.* It is the garnet dodecahedron,
truncated on all the edges. When each edge is
intercepted by two or three small planes, the
terms *bi-emarginated* (bis emarginé), and *tri-emar-
ginated* (tri-emarginé,) are used. Example, Tri-
emarginated garnet grenat tri-emarginé.) which
is the rhomboidal dodecahedron bevelled on all
the edges, and the bevelling edges truncated.

g. *Peri-hexahedral, peri-octahedral, peri-decahedral,*
and *peri-dodecahedral,* when the primitive four-
sided prism is changed by means of decrements
into a six, eight, ten or twelve sided prisms. Crystals in which the primitive form is a regular six-
sided prism, are also named *peri-dodecahedral;*
when the six lateral planes are truncated. Examples, Peri-hexahedral blue vitriol or sulphat
of copper, Pl. lxxii. Fig. 104. HAUY *, which

B b   is

* The Figures here referred to are those in HAUY's System of Mineralogy.

is an oblique four-sided prism, truncated on the obtuse lateral edges : Peri-octahedral blue vitriol or sulphat of copper, Pl lxxii. Fig 105. Hauv, which is the oblique four-sided prism truncated on all the lateral edges : Peri-decahedral blue vitriol or sulphat of copper, Pl. lxxii. Fig 106. Hauy ; the prism truncated on the obtuse lateral edges, and bevelled on the acute lateral edges : And peri-dodecahedral emerald, Fig. 35. *Syst. Min.* which is a six-sided prism truncated on all the lateral edges.

h. *Shortened* (raccourci,) when the primitive form is a prism, whose bases are rhombs, in which the lateral edges contiguous to the great diagonal are intercepted by two planes, so that the primitive form appears to be shortened in the direction of its length. Example, Shortened heavyspar, (baryte sulphatée raccourcie,) Pl. xxxv. Fig. 111. Hauy. It is an oblique four-sided table, very deeply truncated on the acute terminal edges ; or, according to WERNER, a longish six-sided table.

i. *Narrowed,* (retreci,) when the primitive form is a prism, whose bases are rhombs, and in which the lateral edges contiguous to the small diagonal are intercepted by two planes, so that the primitive form appears to be diminished in the direction of its breadth. Example, Narrowed heavyspar, (baryte sulphaté retrecie,) Fig. 142. *Syst. Min.* which is an oblique four-sided table, deeply truncated on the obtuse terminal edges.

2. *Secondary*

2. *Secondary Forms, considered in themselves as being pure-ly Geometrical Forms.*

A Crystal is said to be

a. *Cubical*, (cubique,) when it has the form of the cube, but which in this case is always secondary. Example, Cubical fluor spar.

b. *Cuboidal*, (cuboide,) when the form varies very little from that of the cube, and is very slightly oblique. Example, Cuboidal calcareous spar, (chaux carbonatée cuboide), Pl. XXIII. Fig. 7. HAUY.

c. *Tetrahedral*, (tetraedre,) when the crystal has the regular tetrahedron as a secondary form. Ex-ample, Tetrahedral blende, (zinc sulphuré te-traedre).

d. *Octahedral*, (octaedre,) when it has the octahedron as a secondary form. Example, Octahedral rock-salt

e. *Prismatic*, (prismatique,) when it has the shape of a straight or an oblique prism, in which the late-ral planes are inclined to each other, under angles of 120°. Example, Prismatic calcareous spar, (chaux carbonatée prismatique, Pl. XXIV. Fig 14. HAUY,) and prismatic felspar, (feldspath prismatique,) Fig. 92. *Syst. Min.*

f. *Dodecahedral*, (dodecaedre,) when its surface con-sists of twelve three-sided, four-sided, or five-sided planes, all of which are either equal and si-milar, or differ only in having two kinds of angles. Examples,

a. With

a. With twelve three-sided planes, (the double six-sided pyramid,) viz. rock-crystal, Fig. 67. *Syst. Min.*

b. With twelve rhomboidal or four-sided planes, as in the garnet dodechahedron. Example, Dodecahedral garnet. Fig. 56. *Syst. Min.*

c. With four six-sided and eight four-sided planes, which is the four-sided prism acuminated on both extremities, with four planes, which are set on the lateral edges. Example, Dodecahedral hyacinth. Fig. 6. *Syst. Min.*

d. With twelve five-sided planes; the dodecahedron of Werner. Example, Dodecahedral common iron-pyrites. Fig. 186. *Syst Min.*

g. *Icosahedral*, (icosahedre,) when its surface consists of twenty triangles, of which twelve are isosceles, and eight equilateral. Example, Icosahedral common iron pyrites. Fig 192. *Syst. Min.*

h. *Trapezoidal*, when its surface consists of twenty-four equal and similar trapeziums: it is the double eight-sided pyramid, acuminated on both extremities with four planes, which are set on the alternate lateral edges. Example, Trapezoidal garnet (grenat trapezoidal). Fig. 57. *Syst. Min.*

i. *Tria-contrahedral* (tria-contraèdre), when its surface consists of thirty rhombs; it is the cube, in which each angle is so deeply acuminated with three planes, which are set on the lateral edges, that the lateral planes, and also the acuminating planes, appear as rhombs. Example, Tri-contrahedral

hedral common iron-pyrites, (fer sulphuré tricontaedre). Fig. 194. *Syst. Min.*

k. Enncacontrahedral (ennéacontraèdre), when its surface consists of ninety faces. Example, Enneacontahedral vesuvian (idocrase enneacontraedre), Pl. xlvii. Fig. 74. HAUY.

l. Bi-rhomboidal, when its surface consits of twelve planes, which being taken six and six, and conceived to be elongated until they intersect, afford two different rhomboids : it is, according to the Wernerian view, an acute double three sided pyramid, in which the lateral planes of the one are set on the lateral edges of the other, and acuminated on both extremities by three planes, which are set on the lateral edges. Example, Birhomboidal calcareous spar, (chaux carbonatée bi-rhomboidal), Pl. xxiv. Fig 13. HAUY.

We say in the same sense *tri-rhomboidal;* this, in the Wernerian Crystallography, is a double six-sided pyramid, with alternately broad and narrow lateral planes, in which the broad planes of the one are so set on the narrow planes of the other ; the planes pass beyond the common base, and the pyramid is acuminated on the extremities with three planes, which are set on the smaller lateral planes. Example, Tri-rhomboidal calcareous-spar, (chaux carbonatée tri-rhomboidal). Pl. xxv. Fig. 27. HAUY.

m. Bi-form, tri form (bi forme, tri forme), when it contains a combination of two or three remarkable forms, such as the cube, the rhomboid, the octahedron, the regular six-sided prism, &c.

Example

Example, Tri-form alum (alumine sulphatée tri-forme),Pl. xxxix.Fig 162. Hauy. It is a double four-sided pyramid, deeply truncated on all the edges and angles, in which the truncating planes on the edges originate from the garnet dodecahedron, the truncations on the angles from the cube, and the lateral planes from the octahedron.

n. *Cubo-octahedral, cubo-dodecahedral, cubo-tetrahedral,* when it contains a combination of the two forms indicated by these terms. Examples, Cubo-octahedral fluor-spar, which is the middle crystal between the cube and the octahedron in fluor-spar, Fig. 135. *Syst. Min.* Cubo-dodecahedral common iron-pyrites, Fig. 185. *Syst. Min.* And the cubo-tetrahedral grey copper-ore, which is a simple three-sided pyramid, deeply truncated on all the edges, as Fig. 164. *Syst. Min.*

o. *Trapezian,* when its lateral surfaces consist of trapezia, which lie in two rows, between two bases, as in trapezian heavy-spar (baryte sulphatée trapezienne). Fig. 140. *Syst. Min.* It is a rectangular four-sided table, bevelled on the extremities, where the bevelling planes are trapeziums.

p. *Di-tetrahedral,* that is to say, twice tetrahedral, when it represents a four-sided prism, bevelled on the extremities. Example, Di-tetrahedral tremolite, (grammatite di-tetraedre), Pl. lxx. Fig 214. Hauy.

q. *Di hexahedral* (di-hexahedre), when it is a six-sided prism, having three planes on the extremities.

Example.

Example, Di-hexahedral felspar (feldspath di-
hexaedre), Fig. 95. *Syst. Min.* which is a broad
six-sided prism, bevelled on the extremities, the
bevelling planes set on two opposite lateral edges,
and on each of the extremities, one of the angles,
formed by the meeting of the bevelling planes
with the lateral edges, and on which they are set,
truncated.

In the same sense we say, *di-octahedral, di-decahedral,*
& *di-dodecahedral.* Example, Di-octahedral topaz,
Fig. 30. *Syst. Min.*; di-decahedral felspar ; di dodecahedral asparagus-stone, Fig. 131. *Syst. Min.*
which is a six-sided prism, truncated on the lateral edges, and acuminated on the extremities
with six planes.

r. *Tri-hexahedral, tetra-hexahedral, penta-hexahedral,* and
hepta-hexahedral, (tri-hexaedre, tetra-hexaedre,
penta-hexaedre, hepta-hexaedre), when its surface
consists of three, four, five, or seven ranges of
planes, disposed six and six above each other. Examples, Tri-hexahedral nitrate of potash, Pl. 38.
Fig. 142. HAUY ; which is a six-sided prism,
acuminated on both extremities with six planes :
penta-hexahedral amethyst, Fig. 65. *Syst. Min.*
which is a six sided prism, acuminated on both
extremities with six planes, which are set on the
lateral planes, and the edges between the acuminating and lateral planes truncated : hepta-hexahedral artificial nitrate of potash, Pl. xxxviii.
Fig. 144. HAUY. which is a six-sided prism, a-
cuminated on both extremities with six planes,
which are set on the lateral planes, and the edges
between

between the acuminating and lateral planes be-
vélled.

In the same sense we use the terms *tri-octahedral,*
tri dodecahedral; thus, tri-octahedral sulphat
of lead, Pl lxx. Fig. 76. Haüy, which is a
four sided pyramid, very much elongated, the
edges of the common base truncated, the angles
on it very deeply bevelled, the bevelling planes
set on the lateral edges, and the bevelling edges
again deeply truncated, so that the crystal, viewed in this way, consists of three rows of planes,
of which each row contains eight planes : it
may be more conveniently described as an oblique
four-sided prism : Tri-dodecahedral red silver-
ore, Pl. LXV. Fig. 19. Haüy. It is a six sided
prism, acuminated on the extremities with three
planes, and truncated on all the edges.

s. Bi-geminated (bi-géminé), when it exhibits a combination of four forms, which, taken two and
two, are of the same species, such as the bigemi-
nated calcareous-spar (chaux carbonatée bi-gé-
minée), Pl. xxvii. Fig. 49. Haüy ; which is
an acute double six-sided pyramid, in which the
lateral planes of the one are set obliquely on the
lateral planes of the other, the angles on the
common base truncated, and flatly acuminated
on the extremities with three planes, which are
set on the alternate lateral edges in an unconformable manner, and the edges which the acuminating planes make with the lateral planes
truncated. It is a combination of two rhomboids
and two dodecahedrons.

t. Amphi-

t. Amphi-hexahedral (amphi-hexadre), that is to say,
hexahedral in two senses, because by viewing
the planes in two different directions, we obtain
two six-sided surfaces, such as the amphi-hexahedral axinite (axinite amphi-hexaedre), Pl. 51.
Fig. 107. HAUY, which is a rhomboid truncated in
two opposite acute lateral edges, and also truncated on two of the diagonally opposite edges formed by the meeting of these truncating planes with
the lateral planes

u. Sex-decimal (sex-decimal), when the planes that
belong to the prism or the middle part of it, and
those which belong to the two summits, are the
one six, and the other ten in number, or *vice
versa.* Example, sex-decimal felspar, Pl. 49.
Fig. 86. HAUY, which is a six-sided prism, with
five alterating planes on each extremity.

In the same manner, we say, *octo-decimal, sex-duodecimal,
octo-duodecimal,* and *deci-duodecimal.* Examples,
Octo-decimal artificial blue vitriol, Pl. 73. Fig.
109, HAUY, which is an oblique four-sided
prism, deeply truncated on the obtuse lateral
edges, slightly truncated on the acute lateral
edges, bevelled in the two diagonally opposite
edges formed by the acuter lateral edges with the
terminal planes, and truncated on the other two:
Octo-duodecimal artificial blue vitriol, Pl 73.
Fig. 113. HAUY; the preceding crystallisation,
in which the edges between one bevelling plane
and the terminal plane are again bevelled, so
that there are on this place five small altering
planes: Sex-duodecimal calcareous-spar, Pl. 25.

C c   Fig. 22,

Hauy; which is a very acute double six-sided
pyramid, with unconformable alternately obtuse
and acute lateral edges, the lateral planes of the
one set obliquely on the lateral planes of the
other, and acuminated on both extremities with
three planes, which are set on the acuter lateral
edges : Deci-duodecimal felspar, Fig. 97.

x. Peri-polygonal (peri-polygone), when the prism has
a great number of lateral planes, such as the peri-polygonal tourmaline, Pl. liii. Fig.127. Hauy;
which is a three-sided prism, bevelled on the lateral edges, the bevelling edges truncated, and
the twelve edges formed in this way again truncated.

y. Polysynthetic (surcomposé,) when the form is very
complicated, as in the polysynthetic tourmaline
(tourmaline surcomposé), Pl. liii. Fig. 126.
Hauy; which is a three-sided prism, bevelled
on the terminal edges, the bevelling edges truncated ; and acuminated on the one extremity
with thre planes, on the other with six different
kinds of planes, which together amount to nine-
teen.

z. Anti-enneahedral (antienneadre), when there are
nine planes on the two opposite extremities of
the crystal.   This name belongs to a variety of
tourmaline, Fig. 42. *Syst. Min.* in which there
are nine alterating planes on each extremity, and
the prism has twelve sides, in place of nine, the
usual number.

a a. Prosenneahedral

a a. *Prosenneahedral* (prosenneadre) ; that is to say, having nine faces on two adjacent parts, as in the prosenneahedral tourmaline, Fig. 41. *Syst. Min.* ; in which the prism has nine sides, and one of the extremities nine planes, and the other only three.

b b. *Recurrent* (récurrent), if, on reckoning the planes of the crystal in circular ranges from one end to the other, we have two numbers that succeed each other several times, as 4, 8, 4, 8, 4, as in recurrent tinstone (etain oxydé récurrent), Fig. 252. *Syst. Min.* which may be described as a rectangular four-sided prism, acuminated on the extremities with four planes, which are set on the lateral edges, and the eight edges formed by the acuminating and lateral planes truncated.

c c. *Equidifferent* (équidifferant), when the numbers which designate the faces of the prism, and those of the two extremities, which in this case differ from each other, form the beginning of an arithmetical progression, as 6, 4, 2. Example, Equidifferent basaltic hornblende, Fig. 110. *Syst. Min.* ; which is a six sided prism, acuminated on one extremity with four planes, on the other bevelled.

d d. *Convergent* (convergent), when in the preceding case the series converges rapidly, as 15, 9, 3. Example, Converging tourmaline, Pl. lii. Fig. 124. Hauy ; which is a nine-sided prism, having fifteen planes on the one extremity, and on the other only three.

e e. Unequal

e e. Unequal (impair), when the numbers which designate the planes of the prism, and the planes of
the two summits, which are sensibly different
from each other, are all three unequal, without
forming a progression. Example, Unequal tourmaline (tourmaline impair), Fig. 44. which is a
nine-sided prism, having seven alterating planes
on the one extremity, and three on the other.

f f. Hyper-oxide (hyper-oxyde); that is to say, uncommonly acute, as in the variety of calcareous-
spar, which consists of two rhomboids, of which
the one is acute and inverted, and the other much
more acute, Pl. xxv. Fig. 30. HAUY.

g g. Spheroidal (spheroidal), when its surface consists of forty-eight convex faces, as in the diamond.

h h. Plano-convex (plano-convexe), when the faces are
partly straight and partly uneven, as in the diamond.

3. *Secondary Forms considered in relation to certain
Planes or certain Edges, which are remarkable for
their arrangement or position.*

A Crystal is said to be

a. Alternate, (alterné), when it has upon its upper and
under parts, faces that alternate with each other,
but which correspond on both sides. Example,
Alternate rock-crystal (quartz alterne), which is
a six-sided prism, acuminated on both extremities

ties with six planes, which are alternately small
and large, and conformable.

b. *Bisalternate* (bisalterne), when, as in the preceding
instance, an alternation takes place, not only
among the faces of one and the same part, but
also among those of the two parts. Examples, Bisalternate rock-crystal (quartz bisalterne),
which is the preceding figure, but in which the
larger and smaller acuminating planes alternate
in an unconformable manner; Bisalternate calcareous-spar (chaux carbonatée bisalterne, Pl.
xxv. Fig. 23. Hauy), which is an acute double
six-sided pyramid, with unconformable and alternate obtuse and acute lateral edges, the lateral planes of the one set obliquely on the lateral
planes of the other, so that the edge of the common basis forms a zigzag line; deeply truncated
on the angles of the common base, and in such
a manner, that the acute angles of the trapezoidal
truncating planes, rest on the alternate acute lateral edges, and consequently are alternately
turned towards the upper and under extremity
of the pyramid.

c. *Bibisalternate* (bibisalterne), when there are two
rows of bisalternate planes on each side, as in
the bibisalternate cinnabar, Pl. lxv. Fig. 28.
Hauy.

d. *Annular* or *ring-shaped* (annulaire), when a six-
sided prism has six marginal faces or facets,
disposed in a circular manner around each
base. Examples, Annular emerald (emeraude
annulaire), Fig. 38. *Syst. Min.*; annular tinstone
(étain oxydé annulaire), which is a four-sided
prism, truncated on all the edges and angles,

so

so that it appears like an eight-sided prism
truncated on the terminal edges.

e. *Monostic* (monostique), when a prism with a determinate number of lateral planes, has a row of
facets around each base, differing in number
from those of the lateral planes, and all of which
may be either on the terminal edges, or some on
the terminal edges, and others on the angles.
Example, Monostic topaz (topaz monostique),
Fig. 31. *Syst. Min.* which is a slightly oblique
eight-sided prism, in which two and two lateral
planes meet under very obtuse angles ; slightly
truncated on the four terminal edges, formed by
these lateral planes, and deeply truncated on the
acute angles.

f. *Distic* (distique), when in a similar prism to the
preceding, two rows of facets are arranged around
each base. Example, Distic topaz (topaz distique, Pl. xliv. fig. 41. Hauy, which is the preceding eight-sided prism, in which the terminal
edges, in place of being truncated, are bevelled,
and the angles which the truncating planes of the
acuter angles make with the acuter edges, also
slightly truncated.

g. *Subdistic* (subdistique), when among the facets
which are disposed in the same row around each
base, there are two surmounted by a new facet,
which is as it were the rudiment of a second
row. Example, Subdistic crysolite, (krisolit subdistique), Fig. 121. *Syst. Min.* which is a very
broad four-sided prism, acuminated on both extremities with four planes, which are set on the
lateral planes ; the apex of the acumination
which

which terminates in a line, truncated ; also the la
teral and acuminating edges, and the edges between the small acuminating planes, and the
truncating planes of the acumination.

h. Plagihedral or *diagonal planed* (plagiedre), when it
has facets which are situated obliquely. Examples, Plagihedral or diagonal planed rock-crystal,
(quartz plagiedre), Fig. 69. *Syst. Min.* which is
a six-sided prism, acuminated on both extremities with six planes, which are set on the lateral
planes, truncated on all the angles, and the trapezoidal truncating planes set on obliquely.

i. Unsymmetrical (dissimilaire), when two ranges of
facets situated one above another, on each extremity, exhibit a want of symmetry. Example,
Unsymmetrical topaz (topaze dissimilaire), Pl.
xliv. Fig. 42. HAUY, which is the distic topaz,
with this difference, that the second slight truncations of the angles are wanting; but in place of
them the edges formed by the meeting of the
larger truncating planes of the acute angles with
the bevelling planes, are slightly truncated.

k. Encadré or *framed*, when it has facets which form
kinds of frames or squares around the planes of
a more simple form already existing in the same
species. Example, Framed or squared fluor-spar
(chaux fluatée encadré), Fig. 138. *Syst. Min.*
which is a cube truncated or bevelled on all its
edges.

l. Blunt-edged (prominule), when its edges are very
obtuse. Example, Flat-edged selenite (chaux sulphaté prominulé), Pl. xxxiv. Fig. 99. HAUY.;
which is a twin-crystal, under the form of an
eight-

eight-sided prism, acuminated with four planes,
but of which two and two meet under such obtuse
angles, that the edges are scarcely discernible,
and the acumination has the appearance of a mere
bevelment.

m. *Zoned* (zonaire), when a row of facets is arranged
around the middle part, thus forming a kind of
zone or girdle. Example, Zoned calcareous-spar
(chaux ,carbonatée zonaire), Pl. xxvi. Fig. 39.
Hauy, which is an acute double three-sided pyramid, in which the lateral planes of the one are
set on the lateral edges of the other; the edge of
the common basis truncated, and the angles of
the basis bevelled.

n. *Apophanous* (apophané), when certain faces or
edges offer some indications which assist us in
determining the position of the primitive nucleus,
which otherwise would be detected with difficulty, or even for determining either the direction
or the measure of the decrements. Examples,
Apophanous felspar (feldspath apophané), Pl.
xlix. Fig. 89. which is the didecahedral variety
of felspar, without the truncation and the angles
between the bevelling planes and the lateral edges,
but truncated on the edges of the bevelment.
These last-mentioned truncating planes are inclined to one of the bevelling planes under a less
angle, than to the other, which leads to the remark, that the first bevelling plane belongs to
the primitive nucleus, which otherwise would
have been difficult to determine, on account of
the nearly equal inclination of the two bevelling
planes: Apophanous red silver-ore (argent an-
timonié

timonie sulphuré apophane), Pl. lxiv. Fig. 13.
HAUY, which is a very acute double six-sided
pyramid, with unconformable alternating obtuse
and acute lateral edges; the lateral planes of
the one set obliquely on the lateral planes of the
other, and flatly acuminated on the extremities
with six planes which are set on the lateral
planes. The edge of the common base, and the
acuminating edges, correspond to the edges of
the primitive rhomboid, by which the structure is
revealed to the eye. Apophanous grey copper-
ore (cuivre gris apophané), Pl. lxxv. Fig. 85.
which is a simple three-sided pyramid, bevelled
on all the edges, and acuminated on all the
angles with three planes, which are set on the lateral planes: The bevelling and acuminating
planes point out the decrements.

m. *Blunted* (emoussé), when it has facets which truncate certain parts of the crystal, which otherwise
would be more prominent than the others. Examples, Blunted calcareous-spar (chaux carbona-
té emoussé), Pl. xxvi. Fig. 40. which is the bisalternate variety of calcareous-spar, truncated on
the acuter edges, by which the acute angle of the
truncating planes on the angles of the common
base is obliterated. Blunted axinite, Pl. li Fig.
111. HAUY, which is the amphi-hexahedral variety of axinite, in which the two diagonally opposite edges, which the truncating planes of the
acute edges form with the lateral plane, are bevelled.

n. *Contracted* (contracté), the name given to a dodecahedral variety of calcareous-spar, Pl. xxiv.

D d   Fig.

Fig. 20. Hauy, in which the base of the extreme pentagons experience a kind of contrac-
tion, in consequence of the inclination of the lateral planes. It is a six-sided prism, flatly acuminated on the extremities, with three planes,
which are set on the alternate lateral planes in
an unconformable manner. The lateral planes
towards the ends of the crystal, where they are
without acuminating planes and alternating, are
broader, and somewhat inclined towards each
other, and the acuminating planes are thereby
somewhat diminished in size.

o. *Dilated* (dilaté), the name given to a variety of
dodecahedral calcareous-spar, Pl. xxiv. Fig. 21.
Hauy, in which the bases of the extreme pentagons are in some degree enlarged by the inclination of the lateral planes. It is the preceding
figure in which the lateral planes at the ends
where the acuminating planes rest upon them,
are broader, and incline together, by which the
acuminating planes are shortened.

p. *Acute-angular* (acutanglé), the name given to a prismatic variety of calcareous-spar, Pl. xxvi. Fig. 32.
Hauy, in which the angles are replaced by facets
which form very acute triangles. It is a six-
sided prism, truncated on the angles ; the truncating planes extend far down on the lateral
edges, and are there very acute.

q. *Imperfectly facetted* or *defective* (defective), the name
given to a variety of boracite, in which four of
the angles of the primitive form are obliterated
by facets, whilst the opposite angles remain untouched, so that there arises a certain degree of
imperfection,

imperfection, Pl. xxxiii. Fig. 92. HAUY, which is a cube deeply truncated on all the edges, and on the alternate angles.

r. *Superabundant* (surabondante), the name given to another variety of boracite, where there are four facets, in place of each of the angles which were untouched in the former figures, so that there is a kind of superabundance in place of a deficiency, Pl. xxxiii. Fig. 93. HAUY It is a cube deeply truncated on all the edges, feebly truncated on the alternate angles, but the other four angles are acuminated with three narrow planes, which are set on the lateral planes, and the apices of the acumination are again truncated.

4. *Secondary forms, considered in relation to the Laws of Decrement from which they originate.*

A crystal is named

a. *Unitary* (unitaire), when it experiences only a single decrement by one row. For example, the very acute double six-sided pyramid of sapphire, Fig. 25. *Syst. Min.* If there are three or four decrements by one row, we say *bisunitary, triunitary, quadriunitary.* Examples, Bisunitary, calcareous-spar, Pl. xxiv. Fig. 17 HAUY; Triunitary (triunitaire) chrysolite, Fig. 119. *Syst. Min.*

b. *Binary* (binaire), *bibinary* (bibinaire), *tribinary* (tribinaire), &c. when it experiences one, two, or three decrements by two rows. Examples, Binary calcareous spar, Pl. xxiv. Fig. 11. HAUY. Bibinary calcareous-spar, Pl. xxv. Fig. 26. HAUY.

c. *Ternary,*

c. *Ternary* (ternaire,) *Biternary,* (biternaire,) &c.
when it experiences one, two, or three decrements by three rows.

d. *Unibinary* (unibinaire), when two decrements occur, the one by one row, and the other by two
rows. *Uniternary* (uniternaire), when there is
one by one row, the other by three rows. *Binoternary* (binoternaire), when there is one by two,
and the other by three rows: Examples, Uniternary calcareous-spar, and Binoternary calcareous-spar.

The nomenclature in all the preceding and following expressions has no reference to the planes
which are parallel with those of the primitive
nucleus, which exist most frequently in the secondary crystal Among the forms in which the
nucleus is entirely concealed, some have names
borrowed from different considerations; and
those which remain are so few in number, that
Haüy thought it unnecessary to complicate the
language, by employing a particular designation
for them. In order to avoid confounding together those words that express the decrements
with those that indicate the number of planes;
the former have their termination in *hedral*, as
dodecahedral, or in *al*, as octagonal, whereas the
others end in *ary*.

c. *Equivalent* (equivalent), when the exponent which
expresses one decrement, is equal to the sum of
the exponents of the other decrements.

Example, Equivalent calcareous-spar,

$$\frac{{}^{2}_{c}\,BA}{c\,\overset{1}{g}\,\overset{1}{0}}$$ Pl. xxv. Fig. 28. *Hauy.*

f. *Subtractive*

f. *Subtractive* (soustractif), when the exponent in relation to one decrement is less by one than the
sum of those that indicate the others. Example,

Subtractive calcareous spar, $e \overset{2}{\underset{c}{D}} \overset{2}{\underset{r}{B}}$, Pl. xxvi. f. 37.
$\underset{c}{\overset{}{}} \underset{r}{\overset{}{}} \underset{t}{\overset{3}{}}$

g. *Additive* (additif), when the exponent of the one decrement is greater by one than the sum of the exponents of the others. Example, additive straight

lamellar heavy-spar, $\dfrac{M^5 \; H^{51} \; H^1 \; \overset{2}{A} \overset{1}{E} \; P}{M \quad t \quad\; s \quad d \; o \; P}$

Pl. xxxvi. Fig. 117. HAUY.

h. *Progressive* (progressif), when the exponents form
the beginning of an arithmetical series, as
1, 2, 3. Example, Progressive calcareous-spar,

$\underset{f}{E^1} \; \underset{r}{{}^1E} \; \overset{2}{\underset{m}{D}} \overset{3}{e}$, Pl. xxvii. Fig. 41. HAUY.

i. *Interrupted* or *disjunctive* (disjoint), when the decrements make a sudden spring, as from 1 to 4 or

6. Example, Disjunctive red-silver-ore, $\overset{1}{D}\text{PBB}$,

Pl. lxv. Fig. 22. $\quad\quad\quad\quad n\, P\, \overset{1}{s}\, \overset{4}{c}$

k. *Partial* (partial), when one part remains without
decrements, while other similarly situated parts experience decrements. Example, Partial tin-white

cobalt-ore, $\overset{2}{\underset{e}{B}} \; E^{\frac{1}{2}} \; {}^{\frac{1}{2}}E\underset{M}{M}$, Pl. lxxviii. f. 167. HAUY.

l. *Semidouble* (soudouble), when the exponent which
belongs to one decrement, amounts to half the
sum of the other decrements. Example, Semi-

double topaz, $\dfrac{M^2 \; G^2 \; G^3 \; \overset{2}{B}\overset{2}{E}P}{M \quad u \quad l \quad o\,n\,P}$, Pl. xliv. Fig. 40.

HAUY.

In

In the same manner we say, *Semitriple* (soutriple), *Semiquadruple* (souquadruple), &c.

m. *Doubling* (doublant), *tripling* (triplant), *quadripling* (quadriplant), when one of the exponents is repeated two, three, or four times, in a series which otherwise would have been regular. Examples, Doubling crysolite (peridot doublant).

$$\frac{M^2\ GG^2\ {}^4GG^1\ T\overset{1}{C}\overset{1}{A}\overset{\frac{1}{2}}{B}P}{M\ .\ s\qquad\qquad T\ d\ \mathcal{E}\ k\ P}\quad \text{Fig. 123. } Syst\text{. } Min.$$

Quadrupling crysolite, (peridot quadruplant),

$$\frac{M\overset{1}{G}\ {}^2GG^2\ T\overset{1}{C}\overset{\frac{1}{2}}{A}B\ \overset{1}{B}P}{M\ n\quad s\quad T\ d\ e\ k\quad n\ P},\ \text{Fig. 124. } S.\ M.$$

n. *Identic* or *identical* (identique), when the exponents of two simple decrements are equal to the members of a fraction, which expresses a third and mixed decrement. Example, Identical grey copper-ore,

$$\frac{P\overset{3}{B}BA^2\ A^2\ A^{\frac{2}{3}}\ A^{\frac{2}{3}}}{P\ l\ {}^2\quad o\ \frac{2}{3}\ r},\ \text{Pl. lxxi. Fig. 89.}$$

o. *Isonomous* (isonomé), that is to say, *equality of laws;* when the exponents which indicate the decrements on the edges, are equal to each other, and also those which indicate the decrements in the angles. Example, Isonomous artificial blue vitriol,

$$\frac{{}^1G^1\ M^1\ H^1\ TP^2\ E\overset{1}{I}\overset{1}{G}}{r\ M\ n\ TP\ s\ y\ u},\ \text{Pl. lxxiii. Fig. 108.}$$

Hauy.

p. *Mixed*

p. *Mixed* (mixte), when the form results from a single mixed decrement. Example, Mixed sapphire $\overset{\frac{3}{2}}{\underset{n}{B}}$ Pl. xlii. Fig. 22. Haüy.

q. *Pantogenous* (pantogene), that is to say, which derives its form from all parts of the crystal, when every edge and angle suffers a decrement. Example, Pantogenous heavy-spar,

$$\underset{k \quad M \quad s \quad d \; o \; z \, P}{^1G^1 \; M^1 \; H^1 \; \overset{2 \; 1 \; \frac{4}{3}}{A E B P}}, \text{Pl. xxxvi. f. 118. Haüy.}$$

r. *Biferous* (bifere), when every angle and edge suffers two decrements. Example, Biferous grey copper-ore.

$$\underset{P^1 \; f^3 \; l^u \quad o \quad 1 \quad e}{P \overset{1}{B} \overset{5}{B} B A^2 \; A^2 \; A^{1'}A^1}$$

s. *Surrounded* (entouré), when the decrements occur on all the edges and solid angles around the common basis of a prismatic nucleus. Example Surrounded (entouré) celestine,

$$\underset{M \; z \, o \, d \; P}{\overset{\frac{1}{2} \quad 2}{M B E A B}}, \text{Pl. xxxvi. Fig. 126. Haüy.}$$

t. *Opposite* (opposite), when one decrement is made by one row, and another is intermediate. Example, Opposite tinstone,

$$\underset{M}{M} \left(\underset{z}{\overset{\frac{1}{3}}{A} B^1 \; B^5} \right) B^1, \text{Pl. lxxx. f. 183. Haüy.}$$

u. *Synoptic* (synoptique), when the laws of decrement which occur in all the other Crystals of the same species, or at least in the greater number of them,

them, are united in this crystal. Example, Synoptic felspar.

$$G^2 \ G^4 \ M^2 \ \overset{1 \, 2 \, 3}{H} \overset{1}{T} \overset{1 \, 2 \, 1}{\dot{I} \dot{I} \dot{I}} P \overset{\cdot}{D} \overset{\cdot}{C} I^{\frac{1}{2}} \ I$$
$$l \quad z \quad M \quad z \, T \, x \, y \, q \, P \, n \, n \, o \, o$$
Pl. xlix. fig. 90.

x. *Retrograde* (retrograde), the name given to a variety of calcareous-spar, whose symbol or formu-

la is $\left(\overset{\frac{9}{3}}{\underset{k \, l}{e} \, e} \ \overset{\frac{1}{3}}{\underset{g}{B}} \right)$ contains two mixed decrements,

which are of such a nature, that the resulting
faces seem to retrograde, by throwing themselves
backward on the side of the axis opposite to that
which looks towards the face on which they originate.

y. *Ascending* (ascendant), when all the laws of decrement have an ascending course, in departing from
the angles or lower edges of a rhomboidal nucleus. Example, Ascending calcareous spar,

$\underset{c \, m \, n}{\overset{2 \, 3 \, 4}{e \, e \, D}}$, Pl. xxvii. Fig. 44.

5. *Secondary Forms, considered in relation to their Geometrical Properties.*

A crystal is named

a. *Equiangular* (isogone), when planes occurring on
differently situated places, form among themselves equal angles. Example, Equiangular
chrysoberyl, Pl. xliii. Fig 28. Hauy.

b. *Anamorphic* (anamorphique,) that is to say *inverted shape*, when we cannot give it the most natural position without the nucleus appearing as it
were

were reversed. Example, Anamorphic stilbite or foliated zeolite, Fig. 76. *Syst. Min.* When we view this crystal as an unequiangular six-sided prism, truncated on the angles of the two most obtuse lateral edges, the shape of its nucleus, in comparison with the position it has in the dodecahedral variety of stilbite, appears reversed.

c. *Rhombiferous* (rhombifere,) when certain planes are true rhombs, although, from the manner in which they are cut by the neighbouring planes, they at first sight appear to have no symmetrical figure. Example, the Rhombiferous rock-crystal, Fig. 68. *Syst. Min.* which is a six-sided prism, acuminated on both extremities with six planes, which are set on the lateral planes, and slightly truncated on the alternate angles; the truncating planes are rhombs.

d. *Equiaxe* (equiaxe,) when it has the shape of a rhomboid, in which the axis is equal to that of the primitive rhomboid. Example, the Equiaxe calcareous-spar, Pl. xxiii. Fig. 2. HAÜY, which is a very flat double three-sided pyramid, or it may also be viewed as a very flat rhombus, in which the axis is equal to that of the included nucleus.

e. *Inverse* or *inverted* (inverse,) when it has the form of a rhomboid, the solid angles of which are equal to the plane angles of the primitive rhomboid, and *vice versa.* Example, Inverted calcareous-spar, Pl. xxiii. Fig. 31. HAÜY.

f. *Metastatic* (metastatique,) that is to say *transferred*, when its plane angles and solid angles are

E e   the

the same as those of the nucleus, and are thus
transported to the secondary form. Example, Metastatic calcareous-spar, Pl. xxiii. Fig. 4. HAUY,
which is an acute double six-sided pyramid, with
unconformable and alternating acuter and obtu-
ser lateral edges ; the lateral planes of the one
set obliquely on the lateral planes of the other,
so that the edge of the common base forms a zigzag line. The obtuse angle formed by the edges
of the common base and the acuter lateral edges,
is equal to the obtuse angle of the primitive
rhomboid. The acuter lateral edges are equal to
the lateral edges of the nucleus that lie in one
and the same apex. These two angles, therefore, are as it were transported from the primitive nucleus upon the secondary crystal.

g. *Contrasting* (contrastant,) when it has the form of
a very acute rhomboid, in which there is an in-
version of angles similar to that ,which takes
place in the *inverse*, exhibiting a kind of contrast,
because it in so far resembles in another part
a very obtuse rhomboid. Example, Contrasting
calcareous-spar, Pl. xxiii. Fig. 5. HAUY.

h. *Fixed-angular* (persistant,) is the name of a variety of calcareous-spar, in which certain planes are
so cut by the neighbouring planes, that their angles retain the same magnitude, which they would
otherwise have had, only that the respective positions of these angles are changed. Example, Fixed-angular (persistant) calcareous-spar, Pl. xxv.
Fig. 29. HAUY, which is a six-sided prism very
acutely acuminated on both extremities, with
three planes, which are set on the alternate lateral

ral planes in an unconformable manner, and the
summits of the acuminations deeply truncated.
The angles of the acuminating planes are equal
to the angles of the rhombus of the inverse variety, and the angles of this latter variety have also
remained, notwithstanding the change of form it
has experienced by new or alterating planes.

i. *Analogic* (analogique,) when its form exhibits many remarkable analogies. Example, Analogic
calcareous-spar, Pl. xxvi. Fig. 34. Hauy, which
is a very acute double six-sided pyramid, in
which the lateral planes of the one are set
obliquely on the lateral planes of the other, and
the zigzag edge of the common base so deeply
truncated, that the truncating planes touch each
other, and also the acuminating planes, in a point,
and very flatly acuminated on the extremities
with three planes, which are set on the alternate
lateral edges.

k. *Paradoxical* (paradoxale,) when its structure exhibits very remarkable and unexpected results.
Example, Paradoxical calcareous-spar, Pl. xxvii.
Fig. 42. Hauy, which is the metastatical variety
of calcareous-spar bevelled on the acuter edges,
and acuminated on both extremities with three
planes, which are set on the obtuse lateral
edges.

l. *Complex* (complex,) when its structure is complicated by uncommon laws; as when it is formed
partly by mixed, partly by intermediate decrements. Example, Complex calcareous-spar, Pl.
xxvii. Fig. 43. Hauy, which is the inverse variety,

riety, when we consider it as an acute double three-sided pyramid, bevelled on the edge of the common base, and also truncated on its angles.

6. *Secondary forms, considered in regard to certain particular circumstances.*

A crystal is denominated

a. *Transposed* (transposé,) when it is composed of two halves of an octahedron, or of two portions of another crystal, of which the one appears to be turned upon the other a sixth part of its circumference. Example, Twin-crystal of spinel, Fig. 16. *Syst. Min.*

b. *Hemitrope* (hemitrope,) that is, *one-half turned round,*) when it is composed of two halves of one and the same crystal, of which the one-half appears to be turned round. Example, Twin-crystal of felspar.

c. *Rectangular* (rectangulaire,) the name of a variety of grenatite, which consists of two prisms that intersect each other at right angles, Fig. 63. *Syst. Min,*

d. *Oblique-angular* (obliquangle,) the name applied to the variety of grenatite when two prisms cross each other at an angle of 60°, Fig. 64. *Syst. Min.*

e. *Six-radiated* or *stellular intersecting* (sexradié,) a name given to a variety of grenatite, composed of three prisms, that intersect each another in such a manner, as to represent the six radii of a segular hexagon.

f. Cru-

f. Cruciform (cruciforme,) when it is composed of two crystals which form a kind of cross, as in cross-stone, Fig. 86. *Syst. Min.*

g. Alternately streaked (triglyphé,) when the striæ viewed upon three faces around the same solid angle, are in three directions, and perpendicular to each other. Example, Common iron-pyrites, Pl. lxxvi. Fig. 141. HAÜY.

h. Geniculated (geniculé,) when it is composed of two prisms, which are united at one end, and form a kind of knee. Example, Geniculated rutile, Fig. 218. *Syst. Min.*

IV. *Etraneous External Shape.*

Extraneous External Shapes of Minerals are those derived from organic bodies. They are also named *petrifactions*, and less properly *fossils.* The particular study of these interesting forms belongs to Geognosy, as the oryctognost views them only in a general way. In the prefixed Tabular View, they are arranged in the order in which the originals are described in the natural history of organic bodies, and are first divided into Petrifactions from the Animal Kingdom, and into those from the Vegetable Kingdom *.

A.

---

* In describing petrifactions, with the view of a complete history of the species, a more regular and comprehensive arrangement ought to be followed than that usually employed by naturalists. In a paper which I read before the Wernerian Society some years ago, I proposed and adopted the

following

A. Petrifactions from the Animal Kingdom.

   a. Quadrupeds. The fossil remains of quadrupeds are generally found but little altered, and in single pieces, as *bones, teeth,* and *horns;* seldom in complete skeletons. The greater number of species found in this state appear to be extinct *.

   b. Birds. The remains of birds, which are usually single bones, feet, claws, and bills, are very rare; they have been found in the vicinity of Mont Martre near Paris, and in the limestone of Æningen and Pappenheim.

   c. Amphibious animals. Fossil remains of tortoises and crocodiles have been met with in different parts of Europe. Fossil tortoises occur in the Isle of Shepey in the Medway; and fossil remains of animals allied to the crocodile are met with in the neighbourhood of Bath, in the cliffs

on

---

following arrangement, in describing a petrifaction from Sicily. 1. Description of the external aspect and internal structure. 2. Chemical characters, and chemical composition. 3. Geognostic situation. 4. Geographic situation. 5. Uses. 6. History, under which head I include all that is known of the first discovery of the petrifaction, the names it may have had at different times, the different figures and descriptions of it published by authors, and other information of a miscellaneous nature.

   * It may be remarked, that the fossil remains of the Human Species rarely occur; the only well-authenticated example of this kind being the human skeleton imbedded in an alluvial calcareous mass brought from Guadaloupe by Sir Alexander Cochrane, and now in the British Museum; nor should the rarity of their occurrence excite our wonder, when it is recollected, that human bones are looser in their texture, and more cellular than those of quadrupeds, and therefore much more liable to decomposition.

on the Dorsetshire coast, and also on the coast of Yorkshire.

d. Fishes. Of these we find petrified either the *entire fish, skeletons, vertebræ,* or *teeth.* Of the entire fish instances have been observed in the copper or marl slate of the county of Mansfeld ; and also in Oxfordshire, Gloucestershire, Leicestershire, Lincolnshire, Dorsetshire, and Kent * ; of the skeletons in the limestone of Pappenheim ; of the vertebræ in Shepey ; of the teeth, particularly those of the shark, considerable quantity in the Island of Malta, and also in Kent, and Isle of Shepey.

e. Insects. These are very rare. The only well authenticated instances of petrified fresh water insects are the larvæ of libellulæ found in the limestone of Pappenheim. Of sea insects a very considerable variety have been discovered. Of the genus cancer several distinct species have been found in the Isle of Shepey in the Medway.

Insects inclosed in amber are not to be regarded as petrifactions, because they are dead bodies nearly unaltered.

f. Shells. Many genera of fossil shells are enumerated in the Tabular View, of which a particular account will be given in one of the volumes of my *System of Mineralogy.* It is sufficient for our present purpose to remark, that these fossil remains are uncommonly numerous, and are for the

* Parkinson's Organic Remains, vol. iii. p. 249.

the most part of a species which have never been
found in a living state.

g. *Crustaceous animals.* Of tnese the most remarkable and abundant are the Echinites, and Asterites.

h. *Corals.* Many different fossil genera and species
of these bodies have been figured and described
by naturalists, under the names *madreporites, mil-
leporites, fungites,* &c.

B. Petrifactions from the Vegetable Kingdom.

These are,

a. *Impressions of plants and leaves.* These occur very
frequently, and appear to characterise particular
formations. Thus the impressions and casts of
reeds and ferns appear to occur most frequently
in the bituminous shale and slate-clay of the coal
formation. Petrifactions of seeds and fruits also
occur in sandstone and other rocks.

b. *Transmuted wood,* or *petrified wood.* It occurs in
the form of trunks, branches or roots. The
wood is either petrified with an earthy mineral,
as in wood-stone and wood-opal ; with a metalliferous mineral, as in pyritical wood ; or it is bituminous, as in the different kinds of brown-
coal .

II.

---

* The best English work on Petrifactions is that of Mr Parkinson, entitled " Organic Remains of a Former World." It abounds in curious and
important information, and is adorned and illustrated with numerous
beautiful plates. Mr Sowerby is publishing a useful work entitled *Mineral Conchology.* And the valuable observations on organic remains, in the
Transactions of the Geological Society of London, are further proofs of the
general attention now bestowed on the natural history of petrifactions.

II. *The External Surface.*

The external surface of minerals is either smooth, or more or less uneven. When the inequalities become so great as to affect the shape, they are no longer considered as characterising a variety of surface, but as a variety of external form.

The following are the varieties of this character.

1. *Uneven.* This, of all the kinds of external surface, presents the greatest and most irregular elevations and depressions, yet they are not so considerable as to alter the external shape. Example, Balls of calcedony.

2. *Granulated.* When the surface is composed of numerous small nearly similar roundish elevations, that appear like grains strewed over it, it is said to be granulated. It has a striking resemblance to shagreen. It is either *coarse* or *fine* granulated. The first occurs in reniform brown hematite, also in compact brown ironstone; the latter in diamond grains, and sometimes also in crystals of diamond.

3. *Rough.* This kind of surface is marked with small scarcely visible elevations, which we can hardly discover but by the feel. It has little or no lustre. Examples, Rolled pieces of common quartz and rock-crystal.

4. *Smooth.* Here there is no perceptible inequality, and the surface reflects more light than the preceding kinds of external surface. Examples, Fluor-spar, cubes of galena or lead-glance, and the acuminating planes of rock-crystal.

F f   5

5. *Streaked.* This kind of surface is marked with line-like elevations. It is either *simply streaked* or *doubly streaked.*

A. *Simply streaked,* when the linelike elevations run but in one direction.

a. *Longitudinally streaked.* When the streaks are parallel with the length of the lateral planes. Examples, Topaz, schorl, and beryl.

b. *Transversely streaked.* When the streaks are parallel with the breadth of the lateral planes. Examples, Rock-crystal and quartz.

c. *Diagonally streaked.* Where the streaks are parallel with the diagonal of the planes. We have an example of it in the garnet, where the streaks pass through the obtuse angle of the rhomboid.

d. *Alternately streaked.* When transverse and longitudinal streaks occur on alternate planes. Examples, Cubic iron-pyrites and red iron-stone.

B. *Doubly streaked,* when the streaks run in different directions. This is either

a. *Plumiformly.* When the streaks run obliquely towards a principal streak, like the disposition of the parts of a feather. We must be careful not to confound it with the plumose exter-nal shape. It occurs in the folia of Plumose native bismuth.

b. *Reticularly.* When the streaks either cross each other in a promiscuous manner, or under right angles, forming a kind of flat net-work. It occurs on the surface of silver-white cobalt-ore.

6.

6. *Drusy.* When a crystal is coated with a number of minute crystals of the same kind, so that the new surface acquires a scaly aspect, it is denominated drusy. Examples, Common iron-pyrites and common quartz.

III. *The External Lustre.*

Here we have to consider the *intensity* and the *sort* of lustre.

1. *The intensity of the lustre.* Of this there are five different degrees.

   A. *Splendent.* A fossil is said to be splendent, when in full day light (not in the sunshine) its lustre is visible at a great distance. The highest degree of this is termed *specular splendent.* It generally occurs in minerals with a perfect foliated fracture. Galena or lead-glance, selenite, mica, and iron-pyrites, are good examples of this degree of lustre.

   B. *Shining.* When a mineral at a distance reflects but a weak light, it is said to be shining. Examples, Heavy-spar, pitchstone, and common opal.

   C. *Glistening.* This degree of lustre is only observable when the mineral is near us, and at no greater distance than arm's length. Examples, Grey copper-ore, porcelain jasper, common actynolite, and splintery quartz.

   D. *Glimmering.* If the surface of a mineral, when held near to the eye in full and clear day light, presents a very great number of small faintly shining points, it is said to be glimmering. In strong sunshine it exhibits a kind of play of colour.

lour. As examples of this degree of lustre, we may mention clay ironstone, red hematite, compact galena or lead glance, and porcelain jasper; and of faintly glimmering, Lydian-stone is a good example.

E. *Dull.* When a mineral does not reflect any light, or is entirely destitute of lustre, it is said to be dull. Example, Clay ironstone and chalk.

2. *The sort of lustre.* Of the different sorts of lustre we cannot give any definition, but must rest satisfied with mentioning a few minerals which present these characters in the greatest perfection.

a. *Metallic lustre*, is always combined with opacity. Examples, Copper pyrites, grey copper-ore, and lead-glance.

b. *Semimetallic.* Examples, White and yellowish-grey mica and red hematite.

c. *Adamantine.* It occurs in the diamond, particularly the white and grey varieties, and sometimes also in white lead ore.

d. *Pearly*, as in kyanite, zeolite, and selenite.

e. *Resinous* or *waxy*, as in pitchstone, yellow lead-ore, and tinstone crystals.

f. *Vitreous* or *glassy*, as in rock-crystal and topaz.

In determining the lustre of minerals, we ought to expose them to a strong light, but not to the direct rays of the sun. The specimens should not be handled, a practice too often followed, and which very soon alters the lustre, or adds a lustre to such as have none.

IV. *The Aspect of the Fracture.*

Here we have to observe the *lustre* of the fracture, the *fracture*, and the *shape of the fragments.*

V.

V. *The Lustre of the Fracture.*

The internal lustre, or the lustre of the fracture, presents the same varieties as the external lustre, and therefore requires no particular description.

VI. *The Fracture.*

By fracture we understand the shape of those internal surfaces or planes of a mineral which are produced by breaking or splitting it. These surfaces are either continuous, when the fracture is said to be *compact*, or are composed of a number of line-like or foliated parts, termed *distinct concretions*, when the fracture is named *split* or *divided*.

A. *Compact Fracture.* There are six different kinds of compact fracture, viz. *splintery, even, conchoidal, uneven, earthy*, and *hackly*.

a. *Splintery.* When, on a nearly even surface, small wedge-shaped or scaly parts are to be observed, which adhere by their thicker ends, and allow a little light to pass through, we say that it is splintery. It sometimes passes into even.

b. *Even*, is that kind of fracture-surface which shews the fewest inequalities, and these inequalities are flat and their boundaries never sharply marked, on the contrary, they run into each other imperceptibly. Minerals possessing this kind of fracture have generally a low degree of lustre and of transparency. It occurs in chrysoprase, calcedony, compact galena or lead glance, compact red ironstone, and compact brown ironstone. It passes into large conchoidal and into splintery.

c.

c. *Conchoidal,* is composed of concave and convex roundish elevations and depressions, which are more or less regular: when regular, they are accompanied with concentric ridges, as in many shells, and hence present a conchoidal appearance. It is distinguished, according to the magnitude of the elevations and depressions, into *large conchoidal,* as in obsidian or flint, and into *small conchoidal,* as in pitchstone. The large conchoidal passes into Even, and the small conchoidal into Uneven. It is further distinguished, according to the depth of the inequalities, into *deep conchoidal,* as in rock crystal, and *flat conchoidal,* as in flint: and, lastly, according to perfection, into *perfect conchoidal,* as obsidian, or common opal, and into *imperfect conchoidal,* as porcelain jasper. Minerals 'shewing this kind of fracture exhibit almost every degree of lustre and transparency.

d. *Uneven.* This kind of fracture shews the most considerable elevations and depressions, and the elevations are usually angular and irregular. These elevations are denominated the *grain;* and, according to the size of the grain, the fracture is named *coarse grained,* as in copperpyrites; *small grained,* as in copper nickel; or *fine grained,* as in arsenical pyrites.

This kind of fracture frequently occurs in minerals having some lustre, and which are opaque, and is most frequent in metallic minerals. It passes into small and imperfect conchoidal, and also into earthy.

e. *Earthy.* When the fracture surface shews a great number of very small elevations and depressions, which

which make it appear rough, it is called earthy.
It is always associated with complete opacity
and want of lustre, which latter character dis-
tinguishes it from the fine-grained uneven fracture. It is peculiar to earthy minerals. It is
distinguished into Coarse earthy and Fine earthy.
It passes sometimes into even, and sometimes into
uneven. Examples, Chalk, and clay ironstone.

f. Hackly. When the fracture surface consi ts of numerous small slightly bent sharp inequalities,
which are sometimes only discoverable to the
feel, it is said to be hackly. It occurs only in
native malleable metals, and is, consequently,
accompanied with metallic lustre and opacity.
Examples, Native copper, native silver, &c.

These different kinds of compact fracture often run into each other, and frequently several
occur together ; in the latter case, the most pre-
valent fracture is the one which is to be taken as
the characteristic one.

B. *Split fracture* *. Under this head we include what
is called by some mineralogists the *Structure of
Minerals.*

Three different kinds of split fracture are enumerated in the Tabular View, the *fibrous, radiated,* and *foliated.*

C. *Fibrous fracture.* In this kind of fracture the distinct concretions of which it is composed are
so narrow, that the only magnitude which
can be readily determined, by the naked eye, is
the length ; hence it is be considered as composed

---

* This is the *gespaltenen bruch* of the Germans ; which I have transla-
ted *split fracture,* probably not a very appropriate translation, but I do not
remember any less objectionable.

sed of line-like parts. It is never dull; on the contrary, it is generally glimmering or glistening, seldom shining, and never splendent. It sometimes occurs in transparent minerals, but oftener in those which are nearly opaque. The minerals in which it occurs are sometimes crystallised in capillary crystals. In the fibrous fracture we have to attend to the *thickness*, the *direction*, and the *position* of the fibres.

a. *Thickness* of the fibres.

α. *Coarse fibrous*, when the fibres are of a considerable thickness, as in common fibrous quartz, common asbestus, and fibrous gypsum.

β. *Delicate fibrous*, when the fibres are narrower than in the preceding variety, and occasionally so delicate, as to be scarcely visible to the naked eye. Examples of delicate fibrous fracture occur in red hematite, and fibrous malachite; and of extremely delicate fibrous fracture, in calc-sinter and amianthus.

The coarse fibrous fracture is the link which connects the fibrous with the radiated fracture.

b. The *direction* of the fibres.

α. *Straight fibrous*, as in red hematite and fibrous malachite.

β. *Curved fibrous*, as in asbestus and fibrous gypsum.

c. The *position*.

α. *Parallel fibrous*, when the fibres, whether straight or curved, are parallel to each other, as in common asbestus, and fibrous gypsum.

ε. *Diverging fibrous*, when the fibres proceed from a common centre, in different directions; and this is either

i

   i. *Stellular diverging*, when the fibres diverge in all directions, like the radii of a circle, as in brown hematite.

   ii. *Fascicular* or *scopiform*, when the fibres diverge only on one side, so that the middle fibres are often longer than the lateral ones, as in malachite, fibrous zeolite, and reniform red hematite.

  γ. *Promiscuous fibrous*, when the fibres cross each other in all directions, as in compact plumose antimony.

B. *Radiated fracture.* The distinct concretions in this kind of fracture, have two discernible dimensions, namely, in length and breadth, and of these the first is the most considerable. Hence the fracture surface exhibits long and narrow fracture parts, which sometimes rest on each other, or are placed side by side. The lustre alternates from splendent to shining, and the transparency from translucent to opaque. The minerals in which it occurs are sometimes crystallised either in needles, or in broad prisms. In the radiated fracture, we have to attend to the *breadth*, *direction*, *position*, and *cleavage* of the rays, and the *aspect* of the rays surface.

  a. *The breadth of the rays.*

   α. *Uncommonly broad radiated*, when the breadth of the rays is more than one-fourth of an inch, as is sometimes the case with radiated grey antimony-ore, and kyanite.

   β. *Broad radiated*, when the breadth of the rays is less than the fourth of an inch, but not less

than

than a line, as in common actynolite and mica.

γ. *Narrow radiated*, when the breadth extends from a line to one-fourth of a line, also in actynolite.

b. Direction of the rays.

α. *Straight radiated*, which is very frequent, as in actynolite.

β. *Curved radiated*, which is rare. The curvature is either in the direction of the *breadth*, as in common actynolite, or in the direction of the length, as in kyanite.

c. Position of the rays.

α. *Parallel radiated*, as in grey antimony-ore, and in common hornblende.

β. *Diverging radiated.*

   i. *Stellular*, as in radiated red cobalt-ochre, or cobalt bloom.

   ii. *Scopiform*, as in radiated grey antimony-ore, and radiated zeolite.

γ. *Promiscuous*, as in hornblende slate, and grey antimony-ore.

d. Cleavage or passage of the rays.

a. *Single cleavage* which is generally the case.

b. *Double cleavage*, as in hornblende. In general the cleavage, of which a particular account will be given when treating of the foliated fracture, is imperfect, and we seldom can distinguish more than one variety of it, which is the single.

e. The aspect of the rays surface. The rays are either

α. *Smooth*, as in radiated grey antimony and actynolite.

β. *Streaked*, as in radiated grey manganese-ore and hornblende.

C. *Foliated fracture.* This kind of fracture is composed of folia or planes in which the length and breadth are nearly equal; which are shining or splendent, and superimposed on each other in various directions. It occurs in minerals possessing every degree of transparency, which are generally crystallised, and usually afford regular fragments. It is a more frequent fracture than either the radiated or fibrous.

In the foliated fracture, we have to attend to the *size of the folia,* the *degree of perfection of the foliated fracture,* the *direction of the folia,* the *position of the folia,* the *aspect of the surface of the folia,* and the *passage of the folia,* or *cleavage.*

a. *The size of the folia.* The size of the folia is determined by that of the distinct concretions; so that a mineral which is composed of large granular concretions, must have a large foliated fracture, or, of small granular concretions, a small foliated fracture. When a mineral with a foliated fracture is not composed of distinct concretions, but is one uniform undivided mass, the folia pass uninterruptedly through the whole extent of the mass, and afford the largest variety of foliated fracture.

b. *The degree of perfection of the foliated fracture.* This depends on the facility with which the folia

lia are separated from each other by splitting, and also on the lustre, the smoothness of the fracture surface, and on the readiness with which we can determine the foliated fracture.

Thus it is

α. *Highly perfect*, or *specular splendent :* When the folia are perfectly smooth, and specular splendent, as in galena or lead glance, yellow blende, transparent calcareous-spar, and selenite.

β. *Perfect foliated*, in which the folia are pretty smooth, and shining, and sometimes splendent, as in mica and felspar.

γ. *Imperfect foliated*, when the folia are slightly uneven or even rough, and the lustre lower than in the perfect foliated, as in fluor-spar and beryl.

δ. *Concealed foliated*, when the folia are separated from each other with difficulty, and the foliated fracture appears only in a few places of the fracture surface, as in rock crystal.

c. *The direction of the folia.*

α. *Straight foliated*, as in selenite and calcareous spar.

β. *Curved foliated*, which is either

1. *Spherical curved foliated*, when the folia are so bent, that they resemble either whole spheres or segments of spheres, as in brownspar and mica.

11. *Undulating curved foliated*, when the folia are so laid over each other, that a transverse section gives a serpentine line, but the longitudinal one a straight line, as in mica.

III.

III. *Florifom foliated,* when the folia are variously curved, and the curvatures are arranged
in a scopiform manner, as in galena or lead-
glance.

IV. *Indeterminate curved foliated,* when the folia
are irregularly or indeterminately curved, as
in iron mica or micaceous iron-ore, and
mica.

d. *The position of the folia.*

α. *Common foliated,* when the folia extend
throughout the whole mass, and cover each
other completely, as in calcareous-spar, and
most other minerals with a foliated fracture.

β. *Scaly foliated,* when the folia cover each other
only partially, in their arrangement somewhat resembling the scales on a fish. It is
divided into *large, small,* and *fine* scaly foliated, and occurs in mica.

e. *The aspect of the surface of the folia.* The foliated fracture is either

α. *Smooth,* as in calcareous-spar and felspar:
or,

β. *Streaked,* which is either

I. *Simply streaked,* and in the direction of the
length, as in common hornblende.

II. *Variously streaked,* as in iron-mica.

III. *Plumosely streaked,* as in mica.

f. *The passage of the folia* or *cleavage.*

The cleavage is the number of determinate directions in which a mineral exhibits a foliated
fracture, and according to which it can be split.

It

It is distinguished

α. *According to the number of the cleavages.*

   i. *Single*, when it splits only in one direction, as in mica.

  ii. *Twofold* or *double*, when it splits in two directions, as in felspar, hornblende, and tremolite.

 iii. *Threefold* or *triple*, when it splits in three directions, as in calcareous-spar, rock salt, and galena or lead-glance.

 iv. *Fourfold* or *quadruple*, when it splits in four directions, as in fluor-spar, specular iron-ore, or iron-glance, and beryl.

  v. *Sixfold*, when it splits in six different directions, as in blende and rock crystal.

β. *According to the angle under which the cleavages intersect each other ;* and these exhibit the following varieties,

   i. In the *twofold cleavage*, the two folia or cleavages intersect each other *rectangularly*, as in felspar and hyacinth ; or *oblique angularly*, as in hornblende.

  ii. In the *threefold cleavage*, the folia intersect each other *rectangularly*, as in galena or lead-glance ; *oblique*, yet *equiangularly*, as in calcareous-spar and sparry ironstone ; *oblique* but *unequiangularly*, as in heavy-spar ; and partly *rectangularly*, partly *oblique-angularly*, as in selenite.

 iii. In the *fourfold cleavage*, all the cleavages are *equiangular* and *oblique-angular*, as in fluor-spar, iron glance, and diamond ; or three cleavages are *equiangular* and *oblique-angular*,

lar, in a common axis, and are intersected
by a fourth, which is horizontal and rectangular, as in beryl.

iv. In the *sixfold cleavage*, all the cleavages meet
under *equal oblique angles*, as in rock-crystal; or three of the cleavages are *equiangular* and *oblique-angular* in a common axis,
and are obliquely intersected by three others,
which also intersect the axis in an oblique
direction. Example, hornblende.

These angles of the various cleavages
may also be more particularly measured by
means of the goniometer.

D. *Slaty fracture.* This fracture, like the foliated,
consists of plane-like portions, in which the
length and breadth are nearly alike, but in which
the thickness begins to be discernible. The fracture-surface is generally rough, with but little
lustre. It is nearly allied to the foliated fracture,
but is less perfect, and never occurs in regularly
crystallised minerals, but always in those which
are found in large masses, or in beds. Minerals
with this fracture are generally opaque. This
fracture is further distinguished according to
thickness, direction, perfection, and *cleavage.*

a. *Thickness.*

a. *Thick slaty*, as in alum-slate, flinty-slate, and
clinkstone.

b. *Thin slaty,* as in most of the varieties of clay-
slate.

b. *Direction.*

a. *Straight slaty,* as in common clay-slate.

b. *Curved slaty,* which is either.

     a a. *Indeterminate curved slaty*, as in some va-
         rieties of bituminous marl-slate.
     b b. *Undulating curved slaty*, as in glossy alum-
         slate.
  c. *Perfection.*
     a. *Perfect slaty*, as in clay-slate.
     b. *Imperfect slaty,* as in common flinty slate.
  d. *Cleavage.*
     a. *Single cleavage,* which is the usual variety in
         clay slate.
     b. *Double cleavage,* very rare, as in clay-slate.

2. *Where several fractures occur at the same time, their
relative situation must be observed.*

  A. *One including the other.*

In some minerals there occurs a double fracture, in which the one fracture is larger than the
other, and includes it ; the one, the larger fracture, is named the *fracture in the great ;* the other
the lesser, the *fracture in the small ;* thus, whet-
slate has in the great a slaty-fracture, but in the
small a splintery fracture.

  B. *One traversing the other.*

In other minerals, when the fracture also is
double, but in which the length and breadth are
different, that fracture which is in the direction
of the length is named the *longitudinal fracture :*
the other, in the direction of the breadth, the
transverse or *cross-fracture.* Thus, in topaz,
there is a conchoidal longitudinal fracture, and a
foliated transverse or cross fracture. But in tessular crystals, where the length and breadth are
nearly alike, we use, in place of the term *longi-*
                                    tudinal

tudinal fracture, *principal fracture*, and apply it to
that fracture which occurs the most frequently in
breaking a mineral ; the other fracture, *the cross
fracture*. Thus, in blende, the principal fracture is foliated, with a sixfold cleavage ; the
cross fracture conchoidal ; and in drawing-slate,
the principal fracture is slaty, and the cross fracture is earthy.

VI. *The Shape of the Fragments.*

Fragments are those shapes which are formed when
a mineral is so forcibly struck or split, that masses having
surrounding fracture surfaces are separated from it.

The fragments are either *regular* or *irregular*.

1. *Regular fragments*, are inclosed in a certain number
   of regular planes, that meet under determinate
   angles. They occur only in such minerals as
   have a foliated fracture, with several cleavages.
   Each cleavage in these regular fragments, forms
   two opposite parallel planes, and the shape of
   the fragment depends on the number of these
   planes, and the magnitude of the angles under
   which they meet. Minerals with a twofold cleavage, do not afford perfect regular fragments,
   only prismatic fragments, surrounded with four
   regular lateral planes, as in hornblende   The
   following are the varieties of regular fragments.

   A. *Cubic*, which occur in minerals possessing a rec-
      tangular threefold cleavage, as galena or lead-
      glance and rock-salt.

   B. *Rhomboidal* or *oblique-angular*, which occur in
      minerals having a threefold cleavage, as calca-

H h   reous-spar.

reous spar. When two cleavages intersect each other obliquely, and are intersected rectangularly by a third, the fragments are oblique angular in one direction, and rectangular in another, as in felspar and selenite. In calcareous-spar, the fragments are specular on every side; but in felspar, owing to the imperfect third cleavage, only on four sides.

C. *Trapezoidal.* Occur in foliated coal.

D. *Tetrahedral* or *three-sided pyramidal* and *octahedral*, occur in minerals having a fourfold cleavage, in which the folia meet under equal angles, as in fluor-spar. *Three and six sided prismatic fragments* occur in minerals having a fourfold cleavage, in which three of the cleavages are placed under equal angles around a common axis, and are rectangularly intersected by the fourth, as in beryl.

E. *Dodecahedral.* Fragments of this form occur in minerals having a sixfold cleavage. Sometimes three of the cleavages are disposed around an axis, and are obliquely intersected with other three, as in blende; in other instances all the six cleavages intersect each other under equal hexagon angles, and terminate in an apex, forming *double six-sided pyramidal fragments,* as in rock crystal.

2. *Irregular fragments.*

These have no regular form. They occur in minerals with a single cleavage, and in all the varieties of compact fracture. The following are the different varieties.

A.

A. *Cuneiform*, in which the breadth and thickness
are much less than the length, and gradually
and regularly diminish in magnitude from one
end to the other. It occurs in minerals possessing a scopiform radiated fracture, as Cornish tin ore, red hematite, and radiated zeolite.

B. *Splintery*, in which the breadth and thickness
are less considerable than the length, but without diminution of magnitude from one extremity to the other. It occurs in minerals having parallel fibrous, and radiated fractures,
as in asbestus and bituminous wood.

C. *Tabular*, in which the breadth and length are
more considerable than the thickness, and the
middle is frequently thicker than the sides,
which indeed are sometimes thin and sharp.
It occurs in minerals with a single cleavage, as
mica, also in slaty minerals, as clay-slate, and
there is occasionally a tendency to it in minerals with a conchoidal fracture, as flint.

D. *Indeterminate angular*, in which the length,
breadth, and thickness are in general nearly
alike, but the edges differ much in regard to
sharpness, which gives rise to the following
distinctions.

a. *Very sharp-edged*, as in obsidian and rock
crystal.

b. *Sharp-edged*, as in common quartz, pitchstone and jasper.

c. *Rather sharp-edged*, as in basalt and limestone.

d. *Rather blunt-edged*, as in pumice and copper-
pyrites.

c.

e. Blunt-edged, as in gypsum and steatite.

f. Very blunt-edged, as in fullers-earth and loam.

VII. *The Aspect of the distinct Concretions.*

Distinct concretions are those portions into which certain minerals are naturally divided, and which can be separated from one another without breaking through the solid or fresh part of the mineral. They are separated from one another by natural seams, and frequently lie in different directions. When they are very much grown together, the natural seams are scarcely visible ; in such cases, however, they can be distinguished by their different positions and resplendent lustre. They have been confounded with crystals and fragments, from both of which, as is evident from the preceding definition, they are completely different.

Here we have to consider, 1. The *shape of the distinct concretions.* 2. The *surface of the distinct concretions ;* and, 3. The *lustre of the distinct concretions.*

VIII. *The Shape of the distinct Concretions.*

Distinct concretions, in regard to shape, are distinguished into *granular*, *lamellar*, and *columnar*.

1 *Granular distinct Concretions.*

When the concretions are tessular, or have their length, breadth, and thickness nearly alike, they are said to be granular. It is the most frequent form of the distinct concretion. They are distinguished according to *shape* and *magnitude.*

A. In regard to *shape*, they are

a. *Round granular*, which is either

α. *Spherical*, as in pea-stone and roe-stone.

β. *Lenticular*, as in red granular clay ironstone.

γ.

γ. *Date-shaped*, which is of a longish round shape, as in the quartz near Cullen in Banffshire, and of Prieborn in Silesia.

b. Angulo-granular, which is either

α. *Common angulo granular*, as in galena or lead-glance, and is very frequent.

β *Longish angulo-granular*, as in red hematite and zeolite.

B. In regard to *magnitude*, into

a. Large granular, in which the size exceeds that of a hazel-nut, as in galena or leadglance, blende, and zeolite.

b. Coarse granular, in which the size varies from the size of a hazel nut to that of a pea, as in galena or leadglance, blende, mica, and peastone.

c. Small granular, in which the size varies from that of a pea to that of a millet-seed, as in galena or lead-glance, pea-stone, roe stone, and black blende.

2. *Lamellar distinct Concretions.*

In the lamellar distinct concretions, the length and breadth are nearly alike, and more considerable than the thickness. They occur frequently, but not so often as the granular concretions.

They are distinguished in regard to *direction* and *thickness*.

A. In regard to *direction*, they are

a. Strai ht lamellar, which is either

α. *Quite straight*, as in straight lamellar heavy-spar, or

β. *Fortification-wise bent*, as in amethyst.

L.

 b. Curved lamellar, which is either

  α. Indeterminate curved lamellar, when it is not curved in any particular direction, as in specular iron-ore, or iron-glance.

  β. Reniform curved lamellar, as in red and brown hematite, and native arsenic.

  γ. Concentrical curved lamellar, when they are disposed around a central point. It is divided into *spherical,* as in calcedony and basalt, and *conical,* as in calc-sinter and brown hematite.

B. In regard to *thickness,* into

 a. Very thick lamellar, when the concretions are upwards of half-an-inch thick, as in amethyst, and galena or leadglance.

 b. Thick lamellar, when the thickness varies from half-an-inch to a quarter of an inch.

 c. Thin amellar, when the thickness varies from a quarter of an inch to a line, as in straight lamellar heavy-spar and calcedony.

 d. Very thin lamellar, from that of a line, or the one-twelfth of a line, to the smallest thickness visible to the naked eye, as in straight lamellar heavy spar, native arsenic, and specular iron-ore or iron-glance.

3. *Columnar or Prismatic distinct Concretions.*

In the columnar concretions, the breadth and thickness are inconsiderable in comparison of the length. They are the rarest of the distinct concretions.

They are distinguished in regard to *direction, thickness, shape,* and *position.*

A. In regard to *direction,* they are

    a. Straight columnar, as in schorl, and calcareous spar.

    b. Curved columnar, as in columnar clay iron-stone.

B. In regard to *thickness*, they are

    a. Very thick columnar, when the thickness exceeds half an inch, as in amethyst and prase.

    b. Thick columnar, from half-an inch to a quarter of an inch, as in quartz and calcareous spar.

    c. Thin columnar, from half an inch to the twelfth of an inch, as in columnar clay iron-stone and schorl.

    d. Very thin columnar, when it does not exceed the twelfth of a line, as in schorl. When the concretions become very minute, a transition is formed into the fibrous fracture.

C. In regard to *shape*, they are

    a. Perfect columnar, when the length is considerable, and the thickness uniform from one end to the other, as in calcareous-spar and schorl.

    b. Imperfect columnar, when the concretions are in general short, and sometimes thick in the middle, sometimes at the extremities, as in amethyst and specular iron-ore, or iron-glance. It passes into granular.

    c. Cuneiform columnar, when the concretions become gradually narrower towards one extremity, as in calcareous-spar and quartz.

                              d.

  d. *Ray-shaped columnar*, when the columnar con-
    cretions are compressed, as in specular iron-
    ore or iron glance.   It passes into radia-
    ted.

D. According to the *position*, they are
  a. *Parallel*, as in amethyst.
  b. *Diverging*, as in schorl.
  c. *Promiscuous*, as in calcareous-spar and arseni-
    cal-pyrites.

It may be remarked, that when the concretions occur
very much on the great scale, as is the case in rocks of
the trap formation, a slight alteration of terms is used.
Thus, in place of *granular* we say *massive*, and substitute
tabular for *lamellar*, and always use *columnar*, never *prismatic.*

In several minerals. two varieties of distinct concretions, or different sizes of the same variety, occur together, either the one including the other, or the one traversing the other.   Thus some varieties of schorl are
composed of large granular concretions, and these, again,
are formed of prismatic concretions; some varieties of
straight lamellar heavy-spar, are composed of large granular concretions, and these, again, of thin and straight
lamellar concretions; and peastone affords another example of the same kind of structure, it being composed of
round granular concretions, and each of these of concentric curved lamellar concretions.

In other minerals we observe different kinds of distinct
concretions intersecting each other, as in amethyst,
where curved lamellar concretions intersect prismatic
concretions, and in red and brown hematite, where gra-
                                                        nular

hular concretions are intersected by lamellar concretions.

VIII. *The Surface of the distinct Concretions.*

Distinct concretions exhibit the following varieties of surface.

Smooth, as in hematite and heavy spar; *rough*, as in clay iron-stone; *streaked*, which is either *longitudinally* streaked, as in schorl, *obliquely* streaked as in calcareous-spar, or *transversely* streaked as in amethyst; *uneven* as in brown blende.

IX. *The Lustre of the distinct Concretions.*

It is determined in the same manner as the external lustre.

IV. The General Aspect.

Under this head we include those characters for the sight which are observed in minerals in general   These are, the *Transparency*, the *Streak*, and the *Soiling*.

X. *The Transparency.*

This character presents the five following degrees:

1. When a mineral, either in thick or thin pieces, allows the rays of light to pass through it so completely, that we can clearly distinguish objects placed behind it, it is said to be transparent. It is either *simply transparent*, that is, when the body seen through it appears single, as in mica and

I i   selenite,

selenite; or *duplicating*, when the body seen through it appears double, as in calcareous-spar.

The distance of the two images is in proportion to the thickness of the specimens, and is very inconsiderable in thin pieces. The duplicating property, or double refracting power of calcareous-spar, is observed by looking through two parallel planes; but in some other minerals it is observed by looking through two planes obliquely inclined on each other.

2. *Semi-transparent;* when objects can be discerned only through a thin piece, and then always appear as if seen through a cloud. It is the least frequent variety of this character, and occurs most frequently in siliceous minerals. Examples, Calcedony, common and precious opal, and carnelian.

3. *Translucent.* When the rays of light penetrate into the mineral and illuminate it, but objects cannot be observed either through thick or thin pieces, it is said to be translucent. Examples, Pitchstone, quartz, granular limestone, and massive fluor-spar.

4. *Translucent on the edges.* When light shines through the thinnest edges and corners, or when the edges are illuminated in the same degree as the whole mineral in the immediately preceding variety of transparency, it is said to be translucent on the edges. Examples, Hornstone, heliotrope, and compact limestone.

5. *Opaque.* When even on the thinnest edges of a mineral no light shines through, it is said to be opaque, as in chalk and coal.

The

The Opalescence.

Some minerals, when held in particular directions, re-flect from single spots in their interior a coloured shining lustre, and this is what is understood by opalescence. It is distinguished into

A. *Common* or *Simple Opalescence*, when the lustre appears massive, in undivided rays, as in cat's-eye, and chrysoberyl.

B. *Stellular Opalescence*, when the lustre appears in six rays, or in the form of a star, as in the variety of sapphire named from that circumstance *star-sapphire*. This phenomenon occurs principally in translucent minerals.

XI. *The Streak.*

By the streak, we understand the appearance which minerals exhibit when scratched or rubbed with a hard body, as a knife or steel. In some instances the colour of the mineral is changed; in others the lustre, and frequently neither colour nor lustre are altered.

The *streak*,

    a. In regard to colour, is either

        α. *Similar* to that of the mineral, as in chalk and magnetic cross ironstone; or

        ß. *Dissimilar*, as in specular iron-ore or iron-glance, which has a steel-grey colour, but affords a cherry-red streak; wolfram, which has a greyish-black colour, but a brownish-red streak; and red orpiment, which has an aurora red colour, but affords an orange-yellow streak.

                                        b.

b. *In regard to lustre*, it remains

α. *Unchanged*, as in chalk.

β. *Is increased in intensity*, or *a shining or glist-ening lustre appears in minerals that other-wise have none.* Thus steatite, which is sometimes glimmering, becomes shining in the streak ; and potters clay, fullers earth, and black-brown cobalt-ochre, which have no lustre, become glistening or shining in the streak.

c. *Is diminished in intensity*, or *altogether destroyed.* Thus, grey antimony ore loses its lustre in the streak.

XII. *The Soiling or Colouring.*

When a mineral taken between the fingers, or drawn across another body, leaves some particles, or a trace, it is said to *soil* or *colour*.

It is a character which occurs but in few minerals, and only in those which are soft and very soft. Minerals are said to

1. *Soil*, either

A. *Strongly*, as chalk, drawing-slate, and reddle,

B. *Slightly*, as graphite, or

2. *Do not soil*, as molybdena.

3. *Write*, as chalk. graphite. reddle, molybdena, and black chalk or drawing-slate.

Having now explained the External Characters which are observable by the sight, we proceed to examine those which are made known to us by the senses of touch and hearing.

V.

V. Characters for the Touch.

Here we have to observe, the Hardness, the Tenacity, the Frangibility, the Flexibility, and the Adhesion to the tongue.

XIII. *The Hardness.*

The degrees are

1. *Hard.* When a mineral either does not yield to the knife, or is very slightly affected by it, but affords sparks with steel, it is said to be hard. It is further distinguished according as it is more or less affected by the file.

   A. *Resisting the file,* or *hard in the highest degree,* when it does not yield to the file, but rather acts on it, as diamond, sapphire, and emery.

   B. *Yielding slightly to the file,* or *hard in a high degree,* as in garnet, flint, quartz and calcedony, and very slightly to the knife.

   C. *Yielding readily to the file, but with great difficulty to the knife,* or *hard,* as porcelain jasper, iron pyrites, and felspar.

2. *Semihard.* When a mineral gives no sparks with steel, and yields more readily to the knife than the preceding, it is said to be semihard, as fluorspar, and grey copper-ore.

3. *Soft.* When a mineral is easily cut by the knife, but does not yield to the nail of the finger, it is said to be soft, as calcareous-spar, heavy-spar, serpentine, and galena or leadglance.

4. *Very soft.* A mineral is said to be very soft when it yields easily to the knife, and also

to

to the nail of the finger, as gypsum, steatite and chalk.

In our description of minerals, it is useful to mention their relative hardness, which is ascertained by trying which will scratch the other, by drawing the sharp edge or angle of one on the flat surface of the other. It is, however, of consequence to know, that in crystallised minerals, the solid angles and edges of the primitive forms are very sensibly harder than the angles and edges of the derivative forms, or than the angles and edges produced by fracture, either of crystals or of massive varieties of the same species. This fact has been long known to diamond cutters, who always carefully distinguish between the *hard* and *soft* points of this gem, that is, between the solid angles belonging to the primitive octahedron, and those belonging to any of the modifications; the latter being easily worn down by cutting and rubbing them with the former *. HAUY, in determining the relative hardness, uses plates of calcareous spar, glass, and quartz.

Observations. In examining the hardness of minerals, we must be careful to attend to the following circumstances :

1. Not to confound the real hardness of the mineral with accidental hardness, which latter is caused by the mixture of hard parts in soft minerals, and soft parts in hard minerals.

2.

* Vid. Brückmann's Abhandlung von Edelsteinen, 4ter. Aufl. s. 28, & 29. Mohs uber HAUY's Meionite, in VON MOLL's Efermeriden der Berg und Hüttenkunde 2ter Bandes, 1ste Lief, s. 3. AIKIN's Manual, p. 5.

2. When minerals are composed of distinct concretions, which are not very closely joined together,
we must not give the hardness of the aggregate
for that of the mineral, because the hardness in
such cases must be taken from that of the individual concretions.

3. And we must be careful that the mineral whose
hardness we wish to ascertain, is not in a state of
decomposition.

XIV. *The Tenacity.*

By tenacity is understood the relative mobility or the
different degrees of cohesion of the particles of minerals.
There is a series from the coherent and completely im-
moveable, to the coherent and moderately moveable,
which latter is expressed by malleability, and is the
greatest degree of the mobility of the particles observed
among solid minerals. This series continues through
different kinds of fluid minerals, and the greatest degree
of the mobility of the particles, is found in rock oil. The
degrees of tenacity are,

1. *Brittle.* A mineral is said to be brittle, when on
cutting it with a knife, it emits a grating noise,
and the particles fly away in the form of dust,
and leave a rough surface, which has in general
less lustre than the fracture. In this degree of
tenacity, the particles are completely immove-
able. All hard, and the greater number of semihard minerals are brittle. Examples, Quartz,
heavy-spar, and grey copper-ore.

2. *Sectile* or *mild.* On cutting minerals possessing
this degree of tenacity, the particles lose their
connection in a considerable degree, but this takes
place

place without noise. The particles are coarser than in the brittle variety, and do not fly off, but remain on the knife. The lustre is increased on the streak. This degree of tenacity occurs in most of the soft, and very soft minerals; and the only semihard mineral with this character is native arsenic. Examples, Galena or lead glance, copper-glance, graphite, and molybdena.

3. *Ductile.* Mineral possessing this degree of tenacity can be cut into slices with a knife, and extended under the hammer. The particles are more or less moveable among themselves, without losing their connection. Examples, Native gold, native silver, and native iron.

XV. *The Frangibility.*

By frangibility is understood the resistance which minerals oppose when we attempt to break them into pieces or fragments with a hammer. It must not be confounded with hardness. Quartz is hard, and hornblende soft, yet the latter is much more difficultly frangible than the former. The degrees of frangibility are the following. 1. *Very difficultly frangible,* as in native malleable metals, in silver-glance or vitreous silver-ore, fine granular hornblende, and basalt. 2. *Difficultly frangible,* as hornstone and quartz. 3. *Not particularly difficultly frangible* or *rather easily frangible,* as flint, calcedony and copper-pyrites 4. *Easily frangible,* as opal, calcareous-spar, and fluor-spar. 5. *Very easily frangible,*

as

as straight lamellar heavy spar, galena or lead glance, and slate coal *.

XVI. *The Flexibility.*

This term expresses the property possessed by some minerals of bending without breaking. Flexible minerals are either *elastical flexible*, that is, if when bent they spring back again into their former direction, as mica; or *common flexible*, when they can be bent in different directions without breaking, and remain in the direction in which they have been bent, as molybdena, gypsum, talc, asbestus, and all malleable minerals.

XVII. *The Adhesion to the Tongue.*

This character occurs only in such minerals as possess the property of absorbing moisture, which causes them to adhere to the tongue. It occurs principally in soft and very soft minerals ; it is not known in hard minerals, and there is but one instance of its occurrence in semihard minerals, that is, in the variety of semiopal called *oculus mundi.* The degrees of adhesion are, *strongly adhesive,*

K k   as

---

* Some earthy minerals, such as beryl, flint and opal, when first extracted from their native repositories, are more difficultly frangible than after they have been exposed for some time to the influence of the atmosphere, owing to their containing in these situations a considerable portion of water, which being a nearly incompressible fluid, renders the mineral more difficultly frangible than it is after exposure to the atmosphere, when the water has escaped, and the pores it occupied become filled with air, which is a highly compressible substance.—Vid. AIKIN, p. 9.

as meerschaum, and oculus mundi; *pretty strongly* adhesive, as bole, and potters's-clay ; *feebly adhesive,* as porcelain-earth, chalk, and tripoli ; and *not at all adhesive,* as
quartz and steatite.

VI. CHARACTERS FOR THE HEARING.

XVIII. *The Sound.*

The different kinds of sound occurring in the mineral kingdom are the following : 1. *A ringing sound,* which
is a clear sound, as that of native arsenic, selenite and
rock-crystal.   Specimens to possess this property in full
perfection, should have one or two dimensions, as length
and breadth, greater than the thickness : 2. *A grating
sound,* which is a very weak rough sound, resembling that
emitted by dry wood or fresh burnt clay when rubbed,
and is produced when the finger is drawn across certain
minerals, as mountain-cork and mealy zeolite : 3. *A creaking sound,* which is a harsh sharp sound, as that of natural amalgam, when pressed by the hand.

Having finished the explanation of the characters that
are presented by Solid Minerals, we shall now give an account of those which occur in Friable and Fluid Minerals.   These are very few in number, because few fluid
or friable minerals occur in nature.

PARTI-

II.

PARTICULAR GENERIC EXTERNAL CHARACTERS OF *FRIABLE* MINERALS.

The external characters of Friable Minerals form a particular Section in the System, because they exhibit varieties and kinds that do not occur in solid minerals, and many of the characters of solid minerals, such as fracture, distinct concretions, streak, hardness, and frangibility, and others, are wanting.

I. *The External Shape.*

In Friable Minerals there are but few external shapes. The five following kinds are all that have been hitherto described by naturalists.

1. *Massive,* as in porcelain-earth, and scaly red and brown iron-ores.
2. *Disseminated,* as in earthy azure copper-ore, and blue iron-earth.
3. *Thinly coating* or *incrusting.* It is analogous to the form in membranes. Examples, Copper-black, or black oxide of copper.

4. *Spumous,*

4. *Spumous.* When a friable mineral appears like froth resting on a solid mineral, as is sometimes the case with scaly brown iron-ore.

5. *Dendritic,* also in scaly brown iron-ore.

II. *The Lustre.*

It is determined in the same manner as in solid minerals; we have here, however, the following distinctions.

1. In regard to *Intensity.*
   A. *Glimmering*, which is either strong or feeble, as in scaly brown ironstone, and porcelain-earth.
   B. *Dull*, as in black cobalt ochre.
2. In regard to the *Sort* of Lustre.
   A. *Common glimmering*, as in scaly and brown iron ores.
   B. *Metallic glimmering*, as in scaly red and brown iron ores, and *Pearly glimmering*, as in earthy talc.

III. *The Aspect of the Particles.*

The particles of friable minerals appear in some instances like dust, so that we can with difficulty distinguish by the naked eye any dimensions; these are called *dusty particles*, and occur in cobalt-crust, blue iron earth and porcelain-earth; in others two dimensions can be observed, when they appear foliated, and these are called *scaly particles*, and occur in scaly brown and red iron-ores, earthy talc and chlorite-earth.

IV· *The*

IV. *The Colouring or Soiling.*

Minerals colour either *strongly*, as in scaly red and brown iron ores, and porcelain earth; or *slightly*, as in black cobalt-ochre

V. *The Adhesion to the Tongue.*

This character occurs only in those friable minerals which are cohering. It varies in intensity, being either feeble or strong.

VI. *The Friability.*

Friable minerals are either *loose*, that is, when the particles have no perceptible coherence, as in blue iron-earth; or *cohering*, in which the particles are slightly connected together ; they are either *feebly cohering*, as in porcelain-earth, or *strongly cohering*, as in potters-clay.

PARTI-

III.

PARTICULAR GENERIC EXTERNAL CHARACTERS OF *FLUID* MINERALS.

———

Fluid minerals possess fewer characters than friable minerals. The following four are all that occur.

1. *The lustre* is either *metallic*, as in mercury; or *resinous*, as in rock oil. The lustre is always splendent.

2. *The transparency.* The following are all the degrees necessary for the purposes of discrimination. 1. *Transparent*, as in naphtha. 2. *Troubled* or *turbid*, as in mineral oil: and, 3. *Opaque*, as in mercury.

3. *The fluidity.* Here we have only two degrees to observe, 1. *Fluid*, as in mercury and naphtha; 2. *Viscid*, as is sometimes the case with mineral-oil.

4. *The wetting*, by which we understand the wetting of the fingers when they touch the mineral. It is analogous to the soiling in solid and friable minerals. Mineral oil wets the finger, but mercury does not.

RE-

REMAINING GENERAL GENERIC EXTER-
NAL CHARACTERS.

III. THE UNCTUOSITY.

Some minerals feel *greasy*, others *meagre*, and in order
to distinguish the different degrees of greasiness, the following distinctions are employed.

1. *Very greasy*, as talc and graphite.
2. *Greasy*, as steatite and fullers earth.
3. *Rather greasy*, as asbestus and polished serpentine.
4. *Meagre*, as cobalt.

The greasy and very greasy minerals are generally very soft and sectile, and become shining in the streak.
Mica feels smooth, but not greasy; porcelain earth feels
soft and fine, but not greasy.

IV. THE COLDNESS.

When different kinds of minerals, all having equally
smooth surfaces, are exposed for some time to the same
temperature, we find by feeling them that they possess
different degrees of cold. To use this character with
precision much practice is required; but those who have
accustomed themselves to it, are able, by the mere feel, to
distinguish serpentine, gypsum, porphyry, alabaster,
agate, &c. from one another, and can also distinguish artificial

tificial from true gems. It is, however, principally used
in determining polished specimens. The different degrees mentioned in the Tabular View are,

    1. *Very cold.* Examples, the precious stones, mercu-
       ry, and agate.
    2. *Cold.* Examples, Polished marble or limestone.
    3. *Pretty cold.* Examples, Serpentine, and gypsum
       or alabaster.
    4. *Rather cold.* Examples, Coal and amber.

V. THE WEIGHT.

The degrees of the specific gravity of minerals are the
following.

    1. *Swimming*, or *supernatant*, which comprehends all
       minerals that swim on water, and in which the
       specific gravity is under 1000, water $=1000$.
       Example, Mineral oil, mountain cork, and mine-
       ral agaric.
    2. *Light*, in which the specific gravity varies from 1000
       to 2000. Examples, Amber, sulphur, and black
       coal.
    3. *Not particularly heavy* or *rather heavy*, in which
       the specific gravity varies from 2000 to 4000.
       Examples, Quartz, flint, and calcedony.
    4. *Heavy*, from 4000 to 6000. Examples, Heavy
       spar, copper-pyrites, and iron-pyrites.
    5. *Uncommonly heavy*, all minerals having a specific
       gravity above 6000. Examples, Native metals,
       as gold, silver, &c. ; ores, as galena or leadglance,
       tinstone, &c.

The first and second degrees, which comprehend the
swimming and light minerals, contain all the inflammable
minerals ;

minerals; the third, with a few exceptions, all the earthy minerals; the fourth, the greater number of the ores; and the fifth, the native metals and a few ores.

VI. THE SMELL.

Of this we can give no definition, and shall therefore illustrate it by the minerals in which it occurs.

It is observed either when

1. *Spontaneously emitted,* in which case it is
    a. *Bituminous,* as mineral oil, and mineral pitch.
    b. *Faintly sulphureous,* as natural sulphur.
    c. *Faintly bitter,* as radiated grey antimony-ore.
2. *After breathing on it,* in which a *clayey like smell,* as in hornblende and chlorite, is produced.
3. *Excited by friction.*
    a. *Urinous,* in stinkstone.
    b. *Sulphureous,* in iron-pyrites.
    ç. *Garlick-like,* or *arsenical,* in native arsenic and arsenic pyrites.
    d. *Empyreumatic,* in quartz and rock crystal.

VII. THE TASTE.

This character occurs principally in the saline class, for which it is highly characteristic.

The varieties of it are

1. *Sweetish taste,* common salt.
2. *Sweetish astringent,* natural alum and rock-butter.
3. *Styptic,* blue and green vitriol.
4. *Saltly bitter,* natural Epsom salt.
5. *Saltly cooling,* nitre.
6. *Alkaline,* natural soda.
7. *Urinous,* natural sal-ammoniac

On Describing Minerals.

Most of the species exhibit many varieties of character which are generally distributed throughout a number of individual specimens ; hence it follows, that in order to obtain a distinct conception of a species, we must examine not one, but many different specimens of it. The descriptions of the species ought to be executed in the order laid down and followed in the preceding account of the External Characters; that is, beginning with colour, and then the other characters, in the order there stated.

The following description may serve as an example of the mode of arranging the External Characters.

Precious Garnet.

External Characters.

All the colours of this mineral are *deep-red*, which always inclines to blue ; the principal colour is *columbine-red*, which passes into *cherry-red*, and *blood-red*, and it appears even to run into *brownish red* and *hyacinth-red*.

It rarely occurs massive, sometimes *disseminated*, and in *angular pieces ;* but most frequently in *roundish grains;* and *crystallised*, in the following figures :

1. The *rhomboidal dodecahedron*, which is the fundamental figure.
2. The rhomboidal dodecahedron, *more or less deeply truncated on all the edges.* When the truncating planes become so large as to obliterate the original planes, there is formed the
3. *Leucite crystallisation.*

4.

4. *Rectangular four-sided prism,* acuminated on both extremities with four planes which are set on the lateral edges.

The fundamental crystallisation alternates from *very large* to *very small;* the others are *middle-sized, small,* and *very small.*

The crystals are *all around crystallised,* and *imbedded.*

The *surface* of the grains is usually *rough, uneven,* or *granulated;* that of the crystals is almost always *smooth.*

Internally it alternates from *splendent* nearly to glistening, and the lustre is intermediate between *vitreous* and *resinous.*

The fracture is *more or less perfect conchoidal;* and sometimes *concealed foliated.*

The fragments are indeterminate angular, and rather sharp-edged.

It sometimes occurs in *lamellar distinct contretions.*

It alternates from *transparent* to *translucent.*

It is so hard as to scratch quartz.

It is rather difficultly frangible.

It is *heavy.*

Specific gravity 4.230, *Werner.*

Werner recommends the more essential characters of the mineral to be printed in a different letter from that of the others, in order that they may more readily strike the eye,—a practice which is followed in the preceding description.

MINERAL

MINERAL COLLECTIONS.

In arranging collections of simple minerals, the same method ought to be followed as in the descriptions of them; that is, all the varieties of each character ought to be brought together and arranged in a natural order; and each series should follow in the order of the description. Thus all the different kinds of colour should be ar ranged together, and placed first; next the different varieties of form; then of lustre; fracture; fragments; distinct concretions; transparency; streak, &c. A complete collection, arranged in this way, forms a most interesting and beautiful spectacle, and it also enables us readily to acquire a perfect and distinct conception of the species, however extensive it may be.

The possession of a well arranged Collection of Minerals assists very much in the knowledge and discrimination of mineral species. In the progress of our mineralogical studies, we should omit no opportunity of visiting and examining mineralogical cabinets, the collections of jewellers, the work shops of lapidaries, and in our walks and journies, every mineral that comes in our way ought to be examined and referred to its place in the system. In this island, there are already many valuable mineralogical cabinets, both public and private, the access to which is liberally granted to all those who take an interest in mineralogy.

In Scotland, the most considerable public collections are those of the University of Edinburgh, and of the Professor of Natural History, which are exhibited and
explained

explained during the Lectures on Mineralogy. In the Hunterian Museum at Glasgow there are many valuable and interesting specimens of minerals. The most complete private cabinets in Scotland are those of ROBERT FERGUSON, Esq. of Raith, and of THOMAS ALLAN, Esq. Edinburgh.

The Mineralogical Collection in the British Museum, since the addition of the GREVILLE Cabinet, is now so extensive and complete, that it is considered as one of the most valuable in Europe. The collections of Sir ABRAHAM HUME, and of Mr HEWLAND, are also very interesting, and particularly rich in the rarer and more valuable minerals. The Leskean Cabinet in Dublin, the property of the Dublin Society, originally of considerable extent, is daily increasing, and very lately has received valuable additions by the contributions of their present Professor of Mineralogy, M. GIESEKE *.

CHEMICAL

---

* It is to be regretted, that there are no regular Collections of Minerals to be had in Edinburgh, the lapidaries confining their attention principally to the cutting and polishing of gems, agates, &c. In London, however, the student has an opportunity of purchasing small collections, which will aid him very much in his studies. Mr MAWE, 149 Strand, London, has advertised Collections of this description. On the Continent, many mineral dealers prepare small assortments for the use of the student; the best chosen and arranged collections of this description are those made by the Inspector HOFFMAN at Freyberg, and by Dr LEONHARD at Frankfort on the Maine. More extensive and costly collections are sold by Mr HEWLAND of London, and also by Mr MAWE.

CHEMICAL CHARACTERS

OF

MINERALS.

1. *Action of the Atmosphere.*
2. *Action of Water.*
3. *Action of Acids.*
4. *Action of the Blowpipe.*

CHEMICAL CHARACTERS

OF

MINERALS.

———

THE Chemical Characters of Minerals are those we obtain by their Complete Analysis ; and the changes induced
on them by the action of the Atmosphere,—Water,—
Acids,—and Heat, by means of the blowpipe. In this
Work, we shall not enter on the various modes followed
by chemists in the Complete Analysis of minerals, but
confine ourselves to a short account of the Chemical
Characters obtained by the other means just enumerated.

I. *Action of the Atmosphere.*

Many minerals, on exposure to the atmosphere, experience considerable changes in colour, lustre, hardness, or
decay, fall in pieces, deliquesce, are converted into vitriol, &c. owing partly to the abstraction of the water,
which enters as a constituent part into many minerals,
partly to the absorption of water from the atmosphere

by

by the minerals, or to the oxidation of some of the constituent parts of the mineral.

2. *Action of Water.*

Water either forms a chemical combination with minerals, and completely dissolves them, as is the case with the mineral salts; or it acts by simply destroying their state of aggregation, when the mineral falls into small pieces with an audible noise, as is observed in bole : or it falls without noise, into small pieces, which are soon diffused through the fluid, without either dissolving in it, or becoming plastic, as in fullers earth, and some minerals, as unctuous clays, it renders plastic. Other minerals absorb water in greater or less quantity, which alters their transparency, and also their colour.

3. *The Action of Acids.*

Acids act powerfully on many different minerals, and are the principal agents employed in their complete analysis. When we wish, by means of acids, to obtain some obvious characters, the dilute muriatic acid is that which is generally employed. The native carbonates effervesce, and are soluble in it. In some, as in agaric mineral, calcareous-spar, and witherite, the effervescence is brisk, and the solution rapid; in others, as in dolomite, even when pulverized, the effervescence is feeble, and the solution slow. Some of the earthy minerals which contain silica, water and alkali, in a particular state of combination, if pulverized, and covered with an acid, are, in the space of a few hours, converted into a perfect jelly, as zeolite.

4. *The*

4. The Action of the Blowpipe.

The blowpipe is a tube of silver, copper, brass, or of glass, for delivering a continued stream of air. The stream being directed across a flame, turns it more or less from its vertical position, concentrating it at the same time, and occasioning a more powerful combustion. The air employed is generally either that of the atmosphere, or air which has been breathed; sometimes oxygen gas is made use of, and sometimes an inflammable gas, as the vapour of boiling alcohol. The continued stream of air is furnished by some apparatus, such as a pair of double bellows, a gazometer, a large bladder, or, what is most convenient of all, by blowing with the mouth.

Few persons, Mr Aikin remarks, are able at first, to produce a continued stream of air through the blowpipe, and the attempt often occasions a good deal of fatigue. The first thing to be done is, to acquire the habit of breathing easily, and without fatigue, through the nostrils alone; then to do the same when the mouth is filled and the cheeks inflated with air, the tongue being at the same time slightly raised to the roof of the mouth, in order to obstruct the communication between the mouth and the throat. When this has been acquired, the blowpipe may be put into the mouth, and the confined air expelled through the pipe by means of the muscles of the cheeks; as soon as the air is nearly exhausted, the expiration from the lungs, instead of being made through the nostrils, is to be forced into the cavity of the mouth; the communication is then instantly to be shut again by the tongue, and the remainder of the expiration is to be expired through the nostrils. The second,

second, and all subsequent supplies of air to the blowpipe
are to be introduced in the same manner as the first:
Thus, with a little practice, the power may be obtained
of keeping up a continued blast for a quarter of an hour,
or longer, without inconvenience. Much depends on the
size of the external aperture of the blowpipe. If so
large that the mouth requires very frequent replenishing,
the flame will be wavering, and the operator will soon be
out of breath; if, on the other hand, the aperture be
too small, the muscles of the cheeks must be strongly
contracted, in order to produce a sufficient current, and
pain and great fatigue of the part will soon be the consequence. An aperture, about the size of the smallest
pin-hole, will generally be found the most convenient,
though, for particular purposes, one somewhat larger, or
a little smaller, may be required.

The fuel for the lamp is oil, tallow, or wax ; and of
these the wax is the best, the oil the worst. The wick
should neither be snuffed too high nor too low, and
should be a little bent at its summit *from* the blast of the
pipe. The flame, while acted on by the blowpipe, will
consist of two parts, an outer and inner : the latter will
be of a pale blue colour, converging to a point at the distance of about an inch from the nozzle ; the former will
be of a yellowish colour, and will converge less perfectly.
The most intense heat is just at the point of the blue
flame. The white flame consists of matter in a state of
full combustion, and oxygenates substances immersed in
it : the blue flame consists of matter in a state of imperfect combustion, and therefore partly de-oxygenates metallic oxides which are placed in contact with it.

Various substances are employed for supporting the
mineral, when undergoing the action of the blowpipe.
These

These are of two kinds, combustible and incombustible. The combustible support, used chiefly for ores, is charcoal. The closest grained and soundest pieces of charcoal, of elder or lime tree, are to be selected ; or a support may be made of well pulverised and heated charcoal and gum tragacanth. The gum should be dissolved in water, and powder of charcoal added to it, until it becomes very viscid, when it is to be formed into parallelopipeds, and slowly dried. The incombustible supports are, metal, glass, and earth ; in the use of all which, one general caution may be given, to make them as little bulky as possible. The best metallic support is platina, because it is infusible, and transmits heat to a less distance, and more slowly, than other metals. A pair of slender forceps of brass, pointed with platina, is the best support for non-metallic minerals, which are not very fusible ; for the fusible earthy minerals, and for the infusible ones when fluxes are used, leaf-platina will be found the most convenient. It may be folded like paper into any form, and the result of the experiment may be obtained, simply by unfolding the leaf in which it was wrapped up. Glass supports, are slender glass tubes, on the extremity of which the mineral to be examined is cemented by heating. Earthen supports are, either of small pieces of kyanite, or, when a kind of cupellation is to be performed, they are made of bone-ash, in order to absorb the litharge, and other impurities.

The size of the specimens to be used in our experiments, depends in some measure on the magnitude of the flame to which they are exposed. In a blowpipe having an aperture not larger than a fine pin, the piece ought not to be so large as a pea. A good deal also depends on the fusibility of the mineral ; for if it is very fusible, a much larger piece

may

may be used than when it is difficultly fusible : in the one case, it may be the size of a pea ; in the other, it should not exceed that of a pin's head. The heat first applied to investigate the properties of mineral substances should be very low, not exceeding that which exists a little the outside even of the yellow flame. At this temperature, the phosphorescence is best extricated, and decrepitation for the most part takes place, the fusible inflammables begin to melt, and the metallic and most other mineral salts lose their water of crystallisation. The yellow flame will raise a mineral to a tolerably full red heat ; and it is the temperature best fitted for roasting all the metallic ores. In the still higher degree of heat produced at the point of the interior blue flame, although some minerals still continue refractory, and undergo but little change of any kind, yet the greater part are very sensibly altered. Some, as pearlstone, enlarge very considerably in bulk at the first impression of the heat, but are with difficulty afterwards brought to a state of fusion : others are melted only on the edges and angles ; and in some, a complete fusion takes place.

In examining earthy minerals with the blowpipe, no fluxes are required ; whereas to most of the metallic ores fluxes will be found at almost all times a very useful and often a necessary addition. The ores of the difficultly reducible metals, such as manganese, cobalt, chrome, and titanium, are characterised by the colours which their oxides give to glass. In all these cases, therefore, glassy fluxes must be largely made use of, both to dissolve the earthy matter with which the oxides are generally combined, and to furnish a body, with little or no colour of its own, which may receive, and sufficiently dilute, the inherent colour of the oxide. When the object is not
only

only to dissolve the oxide, but at the same time to retain it at a high state of oxidation, the flux employed should be either nitre, or a mixture of this with glass of borax, or, still better, nitrous borax, formed by dissolving common borax in hot water, neutralizing its excess of alkali by nitrous acid, then evaporating the whole to dryness, and lastly, hastily melting it in a platina crucible. For an active, and at the same time non-alkaline flux, boracic acid may be used, or neutral borate of soda; and where a slight excess of alkali is required, or at least does no harm, common borax by itself, or mixed with a little cream of tartar, when a strong reducing flux is required, may be had recourse to. For coloured glasses, the proper support is leaf-platina, but for reductions charcoal. In the latter case, the ore, previously roasted, if it contain either sulphur or arsenic, is to be pulverised, and accurately mixed with the flux; a drop of water being then added, to make it cohere, it is to be formed into a ball, and deposited in a shallow hole in the charcoal, being also covered with a piece of charcoal, if a high degree of heat is required. In the easily reducible metals, a bit of the ore being placed in the charcoal, and covered with glass of borax, will, in the space of a few seconds, be melted by the blowpipe, and converted into a metallic globule, imbedded in a vitreous scoria. In all cases where a metallic globule is obtained, it should be separated from the adhering scoria, and examined as to its malleability, and other external characters: being then placed a second time on the charcoal, but without flux, it is to be brought to the state of a gentle ebullition, during which, the surface being oxygenated, will exhale a heavy vapour, that condenses on the blowpipe, or falls down on the charcoal, in the form of a powder, or of

specular

specular crystals, from the colour, and other characters of which, the nature of the metal may probably be ascertained. If any suspicion is entertained of a portion of silver or of gold being mixed with the oxydable metal, the button must be placed on an earthen support, and there brought to a full melting heat: by degrees, the oxydable metal will become scorified, and will entirely sink into the support, leaving on the surface a bright bead of *fine metal*, if any such was contained in the alloy ; but the proportion of this last being generally very small, and the entire mass of the alloy often not exc eding alarge shot, it is not unfrequently necessary to have recourse to the magnifying-glass to be fully convinced of the presence or absence of the fine metal *.

As it is of importance to know particularly the various changes induced on minerals by the action of the blowpipe, we shall now give a short enumeration of them †.

I.—*Changes which are effected on Minerals by the simple action of the Heat of the Blowpipe, without the Addition of Fluxes.*

i. Appearances which are not necessarily accompanied with any *permanent change* of the mineral.

a.

---

* The observations on the action of the blowpipe, I owe to Mr AIKIN. —Vid. Man. p. 35.

† Vid, HAUSMANN, in LEONHARD'S Taschenbuch.

a. Phosphorescence. There are two kinds of this phenomenon : in the one there is but a single colour,
which is either white or red ; in the other, there
are many different colours, with a very bright
light, as we observe in fluor-spar and apatite.

b. *The colour of the flame.* Thus, Celestine, or sul-
phate of strontian, colours the blue part of the
flame of the lamp pale red.

ii. Appearances which are associated with *permanent
changes* of the mineral.

A. *Changes which do not affect the form of the mineral.*

1. *Alterations in the Colour,* in which are to be observed,

a. The *tarnish.*

α. *Simple tarnish.*

β. *Variegated tarnish.*   Examples, Pyrites.

b. *Total alteration of colour,* in which the colour
of the whole mass of the mineral is changed.
Thus, Yellow iron-ochre becomes throughout red, red cobalt-ochre blue, and lucullite
loses its colour.

2. *Destruction of the lustre,* of which there are examples in white mica, and foliated gypsum.

3. *Change, or complete destruction of the transparency.* White mica, when exposed to the
flame of the blowpipe, affords an example of
this character.

4. *Change of the refracting power.* Some transparent minerals, when exposed to the blowpipe, become cracked in the direction of the
cleavages, and thus experience a change in

N n   their

their refracting power. This appearance affords us an excellent mean of ascertaining the cleavage of some minerals. Example, Heavy-spar.

5. *Changes in solidity.*

    a. Increase of hardness, as in potters clay.

    b. Calcination or diminution of hardness, as in calcareous spar.

6. *Extrication of smell.*

The mineral named *pyrosmalite* exhales the smell of oxygenated muriatic acid.

7. *Acquiring a taste.*

Thus some, as calcareous-spar, acquire an *alkaline taste* ; others, as heavy-spar, a *hepatic taste* ; and some, as celestine, or sulphat of strontian, an *acid taste.*

B. *Changes which alter the Form, and slightly affect the Substance of the Mineral.*

1. *The loss of water of crystallisation,* as in borax, and alum.

2. *Decrepitation,* or the splitting of minerals into larger and smaller fragments, with more or less force, and with a louder or feebler noise ; by which the water of crystallisation of the mineral is converted into vapour, or the air in its interstices is expanded. Two kinds of this character may be distinguished ; in the one, as in galena or lead-glance, the mineral splits into larger pieces, with a loud noise ; in the other, as in rock salt, it springs into smaller pieces, with a feebler noise.

3.

3. *Sublimation*, when a portion of the mineral is changed into vapour, without undergoing any other change, as is the case with mercury.

4. *Exfoliation*, or the separation of the folia of a mineral from each other, by which its bulk is increased. This is principally caused by the separation of the water of crystallisation. Examples, Foliated gypsum, spodumene, and radiated zeolite.

5. *Efflorescence*, when moss-like shoots appear on the planes or edges of the mineral It occurs, although not in a very striking manner, in topaz. It is probably caused by the escape of a very minute portion of gas.

6. *Intumescence*, when the mineral increases in bulk, during which a number of small air bubbles are formed, thus giving to the mass a spumous aspect. This is caused either by the escape of air or vapour, or both. Examples, Meionite and lepidolite.

7. *Boiling*, when, in melting, a mineral exhibits a bubbling motion, or appears much agitated, as is the case in the fusion of borax, and basaltic hornblende.

8. *Arborisation*, when the whole mass shoots into a fruticose, branched, coralloidal, or contorted form, as in borax, fibrous zeolite, gadolinite, and prehnite.

9. *Rounding*, when the edges and angles are melted and lose thereby their sharpness, as is the case with talc and other minerals.

10.

10. *Glazing*, when only the surface of the mineral melts, and it appears as if covered with a glaze or varnish, as in grenatite.

11. *Fritting*, when single parts of the mass are melted, while others remain unaltered. This character is well seen in those varieties of compact felspar which are mixed with quartz. It often enables us to discover intermixtures in minerals which otherwise would escape our notice

12. *Imperfect fusion*, when the whole mineral runs into an imperfect tenacious mass, and therefore does not form a globule, as in chlorite.

13. *Perfect fusion*, when the whole mineral is perfectly melted, and forms a globule or bead, as in felspar and borax.

14. *Crystallisation*, when a mineral, after it has been fused, and begins to cool, assumes a regular external form. Examples, Brown and green lead-ores, and carbonate of soda.

C. *Changes which alter the Form and Substance of the Mineral.*

1. The *burning* or *oxidation*, when a part, or the whole of the mineral unites with oxygen, during which the following appearances are to be distinguised.

   a. The *glowing*, during which there is a slow combustion, and volatilisation of the combustible, but without flame or smoke, as in glance-coal.

b.

b. The *flaming*. Rapid combustion with flame, in which we have to notice the different colours of the flame, as in black coal.

c. The *smoking*. Volatilisation of the mineral with smoke, which is again readily deposited on cold bodies Examples, Black coal, and native antimony.

d. The *oxidation* or *calcination*, which is the conversion of a mineral into an earthy metallic oxide, either on the surface, or through the whole mass of the mineral.

e. The *vitrification*, or the conversion of the mineral into a glassy metallic oxide.

f. The *coaking*. The conversion of a mineral into a coak, as in black coal.

g. The *incineration*. The conversion of a mineral by combustion into ashes, which occurs in some instances only on the surface, as in glance coal; in others, again, throughout the whole mass, as in brown coal, and some kinds of black coal.

2. The *reduction*, when the mineral is reduced to the metallic state, as white lead ore by the escape of carbonic acid, tinstone by the abstraction of its oxygen, and cinnabar by the escape of its sulphur.

In heating minerals without addition before the blowpipe we have further to attend to the other conditions of the occurring phenomena.

a. The *time* in which they precede each other, whether they follow, α. Very quickly.

β. Quickly.

γ. Slowly.

δ. Very slowly.

b. According as *one* or *several* occur.

   α. When a mineral shews but one of the above mentioned phenomena, as in the melting of compact felspar.

   β. When many occur in succession, as in the bubbling, boiling, and melting of borax.

   γ. When many occur at the same time, as in black coal, when smell, flame, smoke and coaking occur together.

c. The degree of alteration which a mineral experiences, is either

   α. *Great,* when a greater portion of the mass is altered.

   β. *Small,* when a small part of the mineral is changed.

d. *The universality of the change;* in which respect they are said to be

   α. *Complete,* when it refers to all the constituent parts of the whole mineral, as in native antimony, which is entirely volatilised in the form of smoke.

   β. *Partial,* when the change is only in part, not in all the constituents of the mineral. Thus, in antimonial silver-ore, the antimony is volatilised, but the silver remains; and in glance-coal the coaly parts disappears while the earthy parts remain behind in the form of ashes.

II.

II.—*Changes effected on Minerals when mixed with Fluxes, or reducing agents, and exposed to the Heat of the Blowpipe.*

i. *Fluxes.*

Various kinds of fluxes are used in assisting the melting or fusion of the mineral; some of which, under certain circumstances, act as reducing agents, others, as nitre, as oxidising agents. They are first pounded, and those which have water of crystallisation are deprived of it by heating, and either ground fine with the mineral to be examined, and a portion of it, dry or moistened with water, is brought before the blowpipe; or the flux is first brought to a state of complete fusion, and the mineral, either in the state of powder, or in small pieces, is thrown into the liquid flux, and the blowing is continued.

a. The principal fluxes are the following:

1. *Red lead*, for many earthy minerals.
2. *Fluor-spar*, which is a good test for gypsum, with which it forms an enamel.
3. *Gypsum*, which is an excellent test for fluor-spar.
4. *Borax*, which is of very general application in earthy and metalliferous minerals, sometimes as a flux, sometimes as a reducing agent. It is used with most advantage when in the state of glass.

5.

5. *Borax-nitre*, or borax in which its excess of alkali is saturated with nitre.   It is an excellent
flux, particularly for metalliferous minerals.

6. *Carbonate of Soda*, an excellent test for siliceous
minerals.

7. *Cabonate of potash.*

8. *Microcosmic salt.*   A very generally useful flux.

9. *Glass of phosphorus*, which is phosphoric acid in
the state of glass.

10. *Nitre*, a flux particularly fitted for inflammable
minerals, and also a powerful flux for metallic
minerals.

    All the different kinds of flux, with exception
of nitre and borax-nitre, which deflagrate with
charcoal, may be used with any of the usual
supports.

ii *Reducing Agents.*

i. These either abstract oxygen from the mineral, or
protect it from the action of that gas.

Charcoal, which is used as a support, acts in
this way ; but charcoal powder is much more
efficacious.   Oil is sometimes also employed
as a reducing agent, when it is mixed with
the mineral in a finely pounded state.

In treating minerals with fluxes and reducing
agents, we have further to observe,

iii. *Their relation to the Fluxes.*

Here we have to attend to

a. The *solubility* or *insolubility* of minerals in the
fluxes.   During the solution we have to observe, whether it is effected *calmly* ; with
the *evolution of gas*, as in the solution of
grey

grey manganese-ore in borax; or with a kind of *intumescence;* and we have also to notice whether it is effected *quickly* or *slowly.*

b. The *colouring of the flux,* in which is to be noticed,

  α. The *kind of colour* which the melted mass assumes. Thus borax becomes smalt-blue when melted with cobalt-ochre; of a hyacinth-red colour, when melted with a small portion of grey manganese-ore; but of a violet blue colour, when a greater portion of that ore is used; and black with other minerals.

  α. The *degrees of fixity of the colours.* It is in this respect

  1. *Fixed,* as in the blue colours which borax communicates to glass.

  2. *Evanescent,* when the colour remains as long as the mineral is in a state of fusion, but disappears on cooling. Thus glass of borax, in which a ferruginous mineral has been dissolved, has a green colour when in a state of fusion, but which disappears on cooling.

  3. *Changeable,* when the colour of a glass is changed according to the part of the flame in which it is held. Thus glass of borax, with a little oxide of manganese, is of a violet-blue colour when kept in the oxidising part of the flame; but loses this colour when exposed to the reducing part of the flame.

O o

c.

c. The *Reduction of the minerals in the fluxes*, or *reducing agents*, which is either

α. *Perfect*, when all the parts of the mineral are reduced.

β. *Imperfect*, when only part is reduced.

III.— *The various Products obtained by the action of the
Blowpipe on Minerals.*

They may be divided into the following kinds :

A. *Glass*, which is a more or less transparent substance, with a smooth surface, conchoidal fracture, and vitreous lustre. It may be distinguished in regard to

1. *Density*, into

a. *Compact* glass.

b. *Vesicular* glass, with single vesicular cavities.

c. *Spumous* glass, with so many vesicular cavities as to give·the mass a spumous character,
as in obsidian and pitchstone.

2. *Transparency*, into

a. *Clear glass*, which is transparent.

b. *Clouded glass*, with transparent places, but partially clouded.

c. *Muddy glass*, which is only translucent.

B. *Enamel*, which is an opaque body, with smooth
surface, conchoidal fracture, and a vitreous lustre,
sometimes inclining to waxy. It is distinguished in regard to

1. *Density*, into

a. *Compact*, as in gypsum.

b. *Vesicular*, with single vesicles.

c.

    c. Spumous, with many vesicles, as in fibrous
       zeolite.

  2. *Colour,* which is
    Snow-white, as in gypsum, fibrous zeolite and
       gypsum; *greenish-white,* as in nephrite.

C. *Slag,* which is generally an opaque, seldomer a
   translucent body, with the surface full of holes.
   In regard to
  i. *Density,* it is
   a. Compact.
   b. Vesicular.
   c. Spumous.
  ii. *Lustre,* it is
   a. Dull.
   b. Glassy.
   c. Metallic.
  iii. *Colour,* it is
   a. Black.
   b. Brown, &c.
  iv. In regard to the *magnet,* it is
   a. Attracted by the magnet, as in the slag of chlo-
     rite, and several varieties of mica.
   b. Not attracted by the magnet.

D. *Frit,* which is a body exhibiting on its fracture-
   surface parts vitrified and parts unvitrified. Here
   the colour also is to be attended to.

E. *Regulus* or *King,* which is a metallic globule.

F. *Ochre,* an earthy-like metallic oxide, in which the
   colour is to be attended to.

                            G.

G. *Coak,* is common black coal after the dissipation of its more volatile parts.

H. *Charcoal,* is a black, light, feebly glimmering, easily incinerated, *hydro-carbonated* body.

I. *Ashes.*

K. *Incrustation,* which is a fine dust-like powder, deposited during the volatilisation of a mineral on its support, or upon a cold body held over it. It may differ in kind:

    1. *Soot,* a black or brown coaly substance, formed from black coal, &c.

    2. *Sulphur.*

    3. *Metallic oxides,* in which we have to attend to the

        a. *Kind of colour,* which is

            α. White, as in antimonial and arsenical incrustations.

            β. Yellow, as in incrustations of lead.

        b. *Degree of fixity of the colours,* which is

            α. Fixed, as in incrustations of antimony.

            β. Changeable, at different temperatures. Thus bismuth-crust or incrustation continues of a yellow colour, as long as the flame plays upon it; but becomes white on cooling.

IV.—*Changes that take place in the Products after the Experiments are finished.*

Thus heavy-spar may be melted so as to form an enamel; but this enamel is not durable, as it falls to pieces a few hours after it has cooled.

I

Neuman's Blowpipe.

I have just seen, in Mr Brande's *Philosophical Journal*, the description of a blowpipe by Mr Neuman, which appears to be the best hitherto contrived. Mr Neuman's description is as follows.

" Having frequent occasion to condense the air in cavities, I had observed with some surprise the length of time required by the air so confined to escape through such small apertures as might exist, or were purposely made into these cavities, and in conversation with Mr Brooks he suggested, that if the stream were tolerably equable, the principle which gave rise to such an effect might be followed with advantage in the construction of a blowpipe, and I have since verified this idea.

" The instrument I have made consists of a strong plate-copper box perfectly air-tight, three inches in width and height, and four in length, a condensing syringe to force air into the box, and a stop-cock and jet at one end of it to regulate the stream thrown out. The piston-rod of the condenser works through collars of leather in the cap, which has an aperture in the side, and a screw connected with a stop-cock, which may again communicate with a jar, bladder, or gazometer containing oxygen, hydrogen, or other gases. This communication being made, and the condenser being worked, any air that is required may be thrown into the box and propelled through the jet on the flame.

The use of the instrument is very simple. By a few strokes of the piston the air is thrown into the chamber and forms a compressed atmosphere within. When the

cock

cock is opened, the air expanding issues out with great force in a small but rapid stream, which, when directed on the flame of a lamp, acts as the jet from a common blow-pipe, but with more precision and regularity. The force of the stream of air is easily adjusted by opening more or less the small stop-cock, and I have found that with a moderate charge it will remain uniform for twenty minutes; opening the stop cock, or the use of the syringe, will immediately raise it to its first strength.

" These blowpipes are very portable, not liable to injury, and answer, I believe, the expectations of all who have tried them, and I have made many of them for different persons. The whole instrument, with a lamp adapted to it, packs up in a small box not more than six inches in length and four inches in width and height, and there is space enough left for other small articles. I have fitted up boxes rather larger in size with a selection of tests and other useful articles in addition to the blow-pipe, and in this state they form complete mineralogical travelling cabinets."

PHYSICAL

PHYSICAL CHARACTERS

OF

MINERALS.

PHYSICAL CHARACTERS

OF

MINERALS.

_____

Physical Characters, are those derived from physical phenomena, originating from the mutual action of minerals and other bodies. They are highly curious in a general view, but are seldom useful in the discrimination of minerals, as they occur in but few species ; and in these rare cases the same physical properties are met with in very different species. The principal physical characters which occur among minerals, are Electricity, Magnetism, and Phosphorescence.

I. *Electricity.*

It is well known that there are two kinds of electricity ; the one named *positive* or *vitreous*, and the other *negative* or *resinous*.

Electricity can be excited in minerals in three different ways,—by friction, by heating, or by communication with an electrified body. The greater number of minerals which are capable of becoming electrical, acquire

this

this property by friction. Earthy, saline, and metallic minerals, in this way become positively electrified ; whereas, inflammable minerals became negatively electrified. Some minerals by this process become very easily and powerfully electric, while others become electric with difficulty, and exhibit but faint traces of it. A few minerals become electrical by heating, and these belong to the number that also exhibit electrical properties by friction. It has been ascertained, that these minerals have at least two points, of which the one is the seat of positive, and the other that of negative electricity. To these points, which are always placed in two opposite parts of the mineral, HAUY gives the name of *electric poles.* In order to distinguish these poles from each other, the following simple apparatus, figured in Plate VII. is employed. It consists of a needle of silver or copper, having at each end two small balls, *a, b.* This needle, like the common compass-needle, is moveable upon a pivot, or stem, having a very fine point, and at the bottom a broad base or foot. This stem with the needle, are insulated by placing them upon a cylindrical support of resin. To use this apparatus, we place a finger of the left hand on the foot or base of the upright stem, and taking into the right hand a stick of sealing-wax which has been rubbed, present it, during a second or two, at a small distance from the stem ; this being done, we withdraw first the finger, and afterwards the stick. In this way, the needle will be found positively electrified, in such a manner that, according as we approach to one of the balls, the negative or the positive pole of a crystal become electric by heat, the ball will be attracted or repelled. The electricity of the needle will be preserved a quarter of an hour or longer, and we may, while generating it, render it either
ther

ther very sensible or very weak, by varying the distance between the stem and the stick of sealing-wax.

Tourmaline is the mineral in which the property of becoming electric by heat was first observed. It crystallises in prisms, which are frequently nine-sided, and acuminated with three, six or more planes. At the ordinary temperature of the atmosphere it becomes electric by friction, and the electricity which it thus acquires is always positive or vitreous; when two tourmalines are rubbed against each other, the one is electrified positively and the other negatively. But if the tourmaline is heated, it becomes electric; and if its two ends be afterwards approached alternately to the little ball, we shall observe, that the one attracts, and the other repels the ball, from which we may ascertain the poles wherein the respective electricities reside. If one of the poles of the tourmaline be held near light bodies, such as grains of ashes or saw-dust, these minute bodies will be attracted to the stone, and sometimes repelled as soon as they have touched it. When two tourmalines are presented to one another, so that the two positive or the two negative poles are towards each other, they will mutually attract one another; but if two opposite poles are presented to one another, they will mutually repel each other. In order to make this experiment with success, the two crystals should be either balanced on a fine pivot, or suspended by a delicate fibre, or, what is the simplest method, floated upon two pieces of cork. When the two tourmalines are heated, tie one of them upon a flat piece of cork, and present to one of its poles the two poles of another tourmaline in succession. When two similar poles are towards each other, the floating tourmaline will turn round and present the opposite pole; and when two
opposite

opposite poles are presented to each other, the floating tourmaline will follow the other in all its motions, just like a floating-needle guided by the action of a magnet.

Tourmaline exhibits electrical properties by exposure to heats from $99\frac{1}{2}°$ to $212°$ of Fahrenheit. When, however, it is more and more heated, there is a term when it ceases to exhibit signs of electricity. It often happens, that after withdrawing it from the fire, we are obliged to leave it to return to a moderate temperature before it exhibits any action upon light bodies presented to it. " It would seem, (says Haüy,) that beyond the term where its electricity has become insensible through the action of too strong a heat, there is another where its effects are reproduced in an inverse sense. We have caused the foci of two burning-glasses to fall upon the extremities of a tourmaline, and have observed that each pole, after having acquired its ordinary electricity, would next cease to act, and, lastly, would pass to the opposite state ; so that the attraction, after having become zero, would give place to repulsion, or reciprocally."

If a tourmaline be broken when in a state of excitation by heat, each fragment, however small it may be, has two opposite poles,—a phenomenon analogous to what takes place in a broken magnet.

In the tourmaline, the electric density diminishes rapidly from the summits or poles towards the middle of the crystal, and is almost nothing throughout a sensible space towards the middle of the prism. The greatest density which resides in the positive and negative poles, is near the summits. This distribution is analogous to that of the electric fluid diffused about a cylinder. It

may

may be rendered perceptible to a certain degree, by
moving a tourmaline backwards and forwards, that has
one of its planes opposite one of the balls of the little
needle; we shall observe, that this ball has a marked tendency towards one point of the mineral; but when the
needle points to the middle of the prism, so that it is
equidistant from the two poles, the needle will have no
motion except a mere fluttering given to the ball.

Hauy has ascertained, that the polarity which the
tourmaline and other minerals receive from heat, is rela-
ted to the form of their secondary crystals. The opposite and corresponding ends of crystals are generally similar, both with regard to the number, to the disposition,
and the figure of their planes. The forms of crystals,
however, that become electrical by a change of temperature, deviate from this symmetry of form; so that the
poles, or parts of the crystal where the opposite electricities are situated, although they are similarly situated at
the two extremities of the secondary crystal, yet differ in their configuration; one of them undergoing decrements which are evanescent on the opposite end, or to
which decrements correspond that are subjected to another law,—a circumstance which may enable an observer
to predict beforehand, simply from the inspection of the
crystal, on which side either species of electricity will be
found, when the crystal shall be submitted to the test of
experiment. Thus, in the variety of tourmaline called
isogone, which is a nine-sided prism acuminated on one
extremity with three planes, and on the other with six
planes, experiments prove, that the first summit is the
seat of resinous electricity, while the second manifests
vitreous electricity.

The

The same deviation from the rules of symmetry in the secondary form of crystals, has been observed in the octosexdecimal topaz.

But of all the crystals that exhibit this co-relation between the exterior configuration, and the electric agency, the most remarkable are those which appertain to the boracite, whose form is, generally, that of a cube truncated on all its edges, and also on the angles. Here the two electricities act according to the direction of four axes, each of which passes through two opposite solid angles of the cube, which is the primitive form. In one of the varieties named *defective boracite,* one of the two solid angles situated at the extremities of the same axis, is entire; the other is truncated. Now negative or resinous electricity is evinced at the angle which has not undergone any alteration; and positive or vitreous electricity at the truncated angle; thus making eight electric poles, four for each species of electricity.

" We may now ask," says Hauy, " whether, in the midst of the imposing apparatus of our artificial machines, and of that diversity of phenomena which it presents to the astonished eye, there is any thing more calculated to excite the interest of philosophers, than these little electrical instruments executed by crystallisation, than this combination of distinct and contrary actions, confined within a crystal whose greatest dimension is probably less than the twelfth of an inch ? And here the observation we have so often previously made recurs to the mind with additional force, that those productions of nature which seem desirous to conceal themselves from our notice, are they which may reward us most liberally

for

for a closer examination." HAUY's *Natural Philosophy*, from p. 425 to 435, vol. i.

The third mode of exciting electricity in minerals, or that by communication, occurs only in minerals which are in a pure metallic state.

II. *Magnetism.*

Very few minerals are magnetic; it is a character which occurs principally in ores of iron, or in such minerals as contain a portion of metallic iron, or iron in the state of black oxide.

A good many minerals, after exposure to the blowpipe, become magnetical.

III. *Phosphorescence.*

Some minerals, when rubbed or heated, emit in the dark a more or less shining light, or are said to be phosphorescent. Thus, yellow blende, when scratched with a hard body, emits a strong light. When two pieces of quartz are forcibly struck against each other, both become luminous; and fluor-spar, when heated, becomes phosphorescent.

GEO-

GEOGNOSTICAL AND GEOGRAPHICAL CHARACTERS.

These characters do not require any particular consideration.

FINIS.

PLATE I.

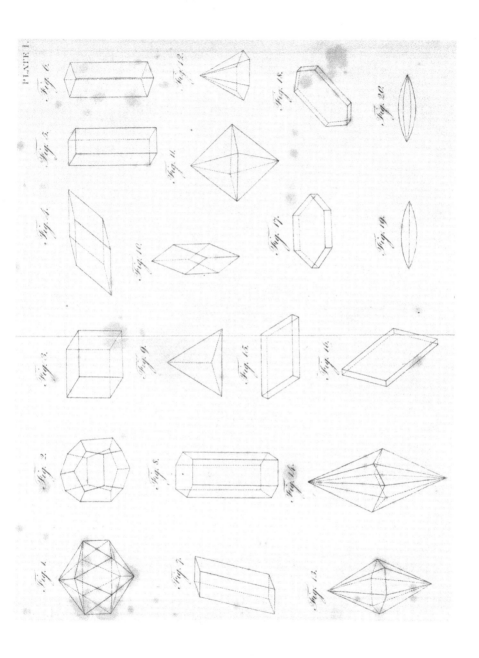

Fig. 1. Fig. 2. Fig. 3. Fig. 4. Fig. 5. Fig. 6. Fig. 7. Fig. 8. Fig. 9. Fig. 10. Fig. 11. Fig. 12. Fig. 13. Fig. 14. Fig. 15. Fig. 16. Fig. 17. Fig. 18. Fig. 19. Fig. 20.

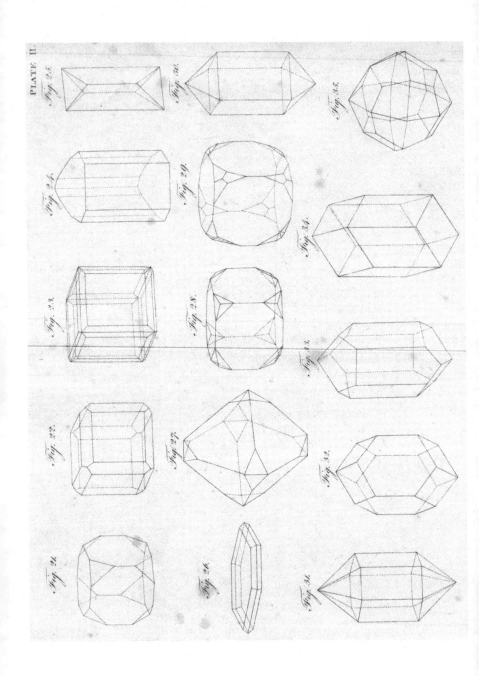

PLATE II.

Fig. 25. Fig. 30. Fig. 35. Fig. 26. Fig. 29. Fig. 34. Fig. 33. Fig. 28. Fig. 5. Fig. 22. Fig. 27. Fig. 32. Fig. 24. Fig. 23. Fig. 31.

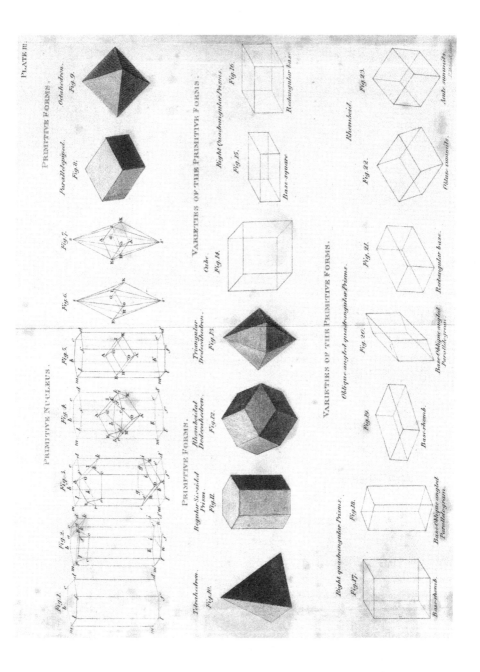

The material originally positioned here is too large for reproduction in this reissue. A PDF can be downloaded from the web address given on page iv of this book, by clicking on 'Resources Available'.

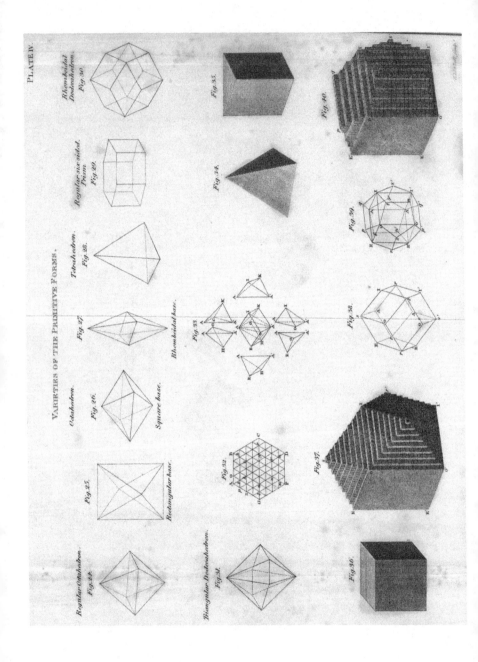

PLATE IV.

VARIETIES OF THE PRIMITIVE FORMS.

Rhomboidal Dodecahedron. Fig. 40.

Regular six-sided Prism. Fig. 29.

Tetrahedron. Fig. 28.

Octahedron. Fig. 27.

Fig. 25.

Regular Octahedron. Fig. 24.

Triangular Dodecahedron. Fig. 31.

Fig. 33.

Fig. 34.

Fig. 41.

Fig. 40.

Fig. 39.

Fig. 38.

Fig. 37.

Fig. 36.

Fig. 32.

Rhomboidal base.

Square base.

Rectangular base.

The material originally positioned here is too large for reproduction in this reissue. A PDF can be downloaded from the web address given on page iv of this book, by clicking on 'Resources Available'.

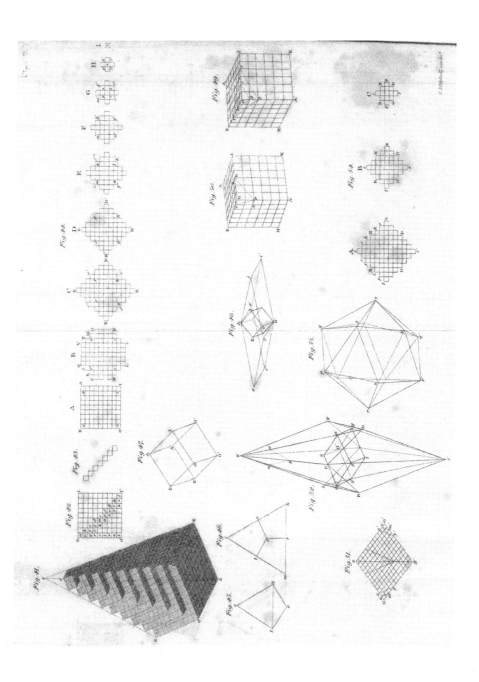

The material originally positioned here is too large for reproduction in this reissue. A PDF can be downloaded from the web address given on page iv of this book, by clicking on 'Resources Available'.

PLATE VI

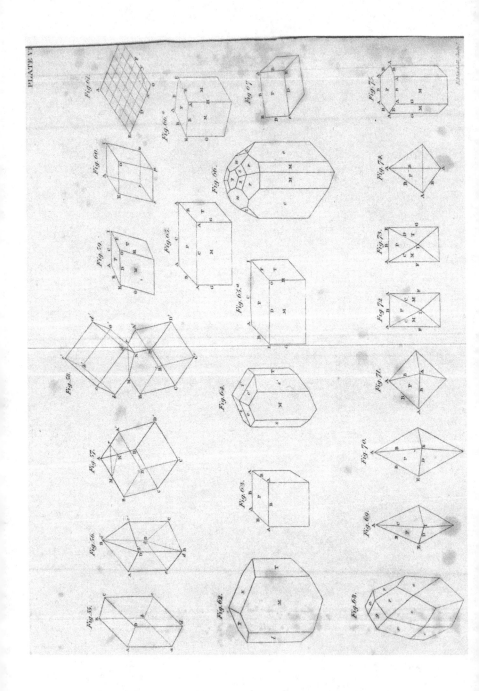

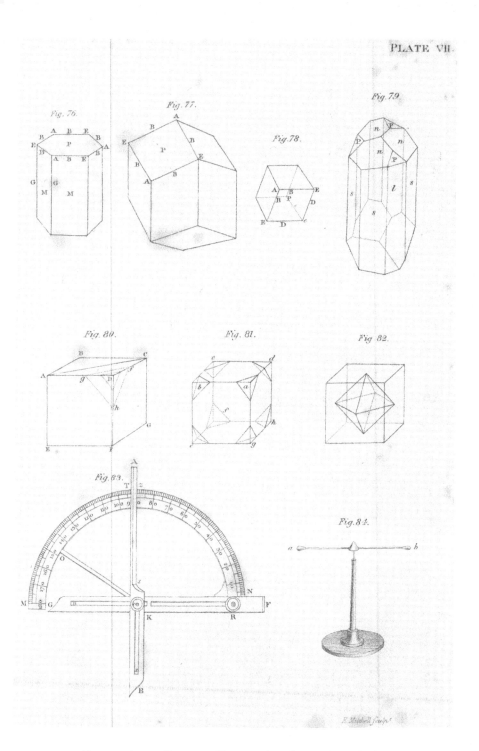

PLATE VII.

Fig. 76.   Fig. 77.   Fig. 78.   Fig. 79.

Fig. 80.   Fig. 81.   Fig. 82.

Fig. 83.   Fig. 84.

E. Mitchell sculp.t

Printed in the United States
By Bookmasters